# ROBOTS
## Planning and Implementation

Chris Morgan

IFS (Publications) Ltd., UK

Springer-Verlag
Berlin Heidelberg New York Tokyo
1984

**British Library Cataloguing Publication Data**

**Morgan, Chris**
**Robots.**
1. Robots, Industrial
I. Title
629.8'92 TS191

ISBN 0-903 608-36-7 IFS (Publications) Ltd.
ISBN 3-540-125841 Springer-Verlag Berlin Heidelberg New York Tokyo
ISBN 0-387-125841 New York Heidelberg Berlin Tokyo



Typesetting by Gilbert Composing Services, Leighton Buzzard, UK
Printed by Butler and Tanner Ltd., Frome, UK

To my wife Pamela and my children
Damian, Ruth, Claire and Benedict

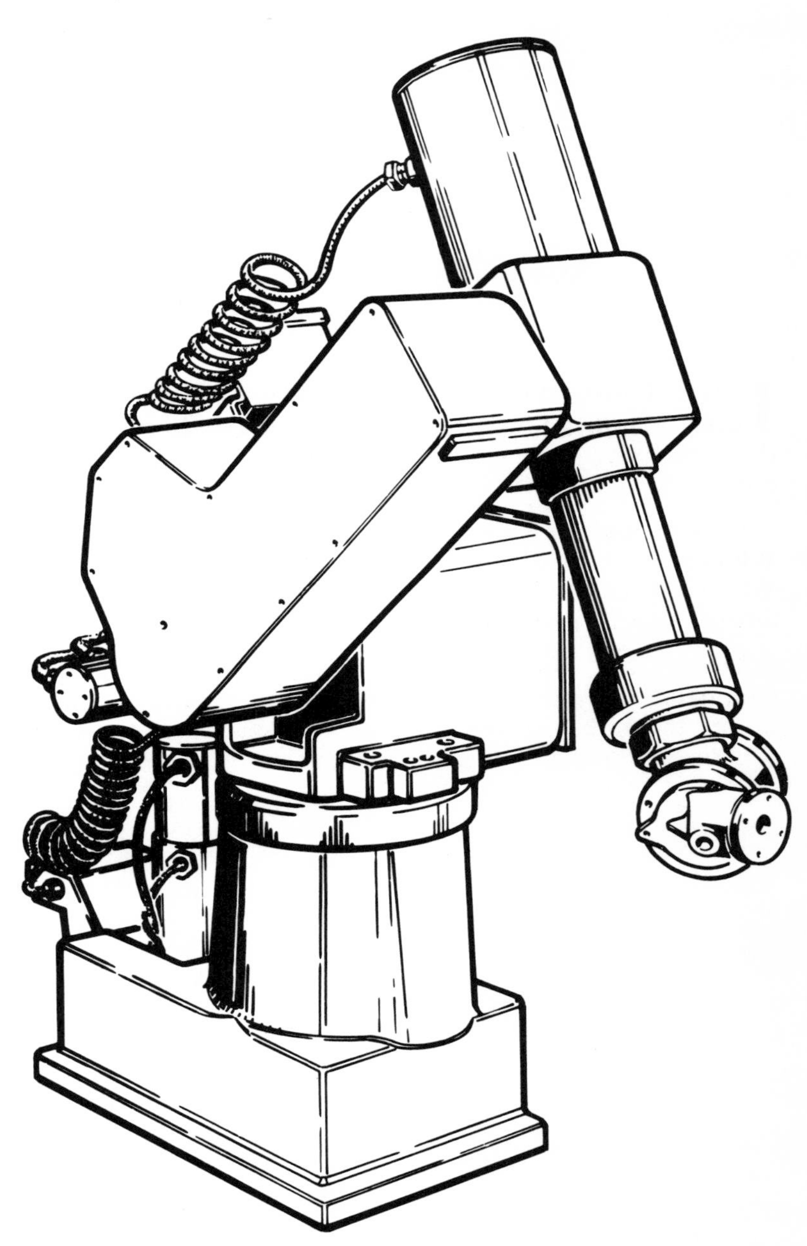

A robot may be distinguished from other types of automation by the fact that it can be programmed and reprogrammed to suit the varying demands of production as and when they occur

# Acknowledgements

I WISH to express my gratitude to the people and organisations who have helped me during the preparation of the book by providing information and by giving constructive criticism of its contents.

In particular I wish to express my deep appreciation to the Whitworth Foundation whose generous award of the Whitworth Exhibition enabled me to travel abroad to collect information for the case studies.

Thanks are also due to Professor Keith Rathmill of Cranfield Institute of Technology for his help and comments during the writing of the book, and to Michael Innes of IFS (Publications) Ltd. for his careful and thoughtful editing.

# Foreword

Most of the recent books on Robotics have been written for the specialist and rarely understood by anyone just trying to get some advice on installing a few robots. The reason for this trend is quite understandable, because of the very rapid advances in robotic and computer technologies.

It is therefore an important occasion when a practical book, based on experience, is launched on to the market. This book will appeal to both the young engineer studying at college or university and the manager trying to bring himself up to date in this new and confusing technology. There are also lessons to be learned by many of us that tend to over specialise as a result of our technical inclinations, and to the detriment of the cost implications.

*Planning and Implementation* is an unpretentious title for a book that is far more than merely informative; readers who seek practical explanations on how to go about installing robotic systems are treated to clear and accurate case studies of how it can be achieved.

This book should therefore hold a prestigious place in the technical libraries of the world.

I wish it every success.

David Teale
IBM UK Laboratories Ltd.

# Contents

## *Part III*
## *Case Studies*

# Preface

ONE OF the exciting aspects of working with robots and their related technologies is the constant progress that is being made. It would be impossible to finish writing a book of this nature if all of these new developments were to be included. Instead I have concentrated on some of the basic aspects of applying one of the key elements in the majority of modern manufacturing systems—robots.

It is a fact that robots have not been applied in industry to the extent that they are commonplace. Because of this there is no established body of theory for reference. Fortunately this is beginning to change as the study of robots is now becoming more commonplace in universities and polytechnics throughout the world. With this in mind it is clear that there remains many grey areas of robot application which are subject to interpretation in the light of personal experience. This book reflects my own experience in the design and implementation of fourteen robot systems in a variety of applications.

I have written this book with the first time user in mind. It will act as a guide as to what should be done and in what sequence. Some details have inevitably been omitted as they will only be settled with respect to any particular application. However, with these reservations in mind the book will be useful to both manufacturing management and students of robotics or manufacturing sciences.

# Introduction

IN THE developed countries of the world there is an ever present need to maintain economic growth. Without growth there is stagnation and decline. Historically growth has been realised on a manufacturing base that has harnessed abundant raw materials and a plentiful supply of labour. Times change, and the rate of economic growth has tended to moderate.

Traditionally abundant supplies of raw materials have diminished and growth for the future is tied to the efficient use of new technologies and processes that can convert new raw materials to supply the world's needs. The differences between these new technologies and processes and the old, lies largely in the utilisation of labour. The days of the unskilled labourer are passing and tomorrow's world will be built using highly skilled technicians, engineers and scientists.

This underlying trend is already in evidence and one has only to look at the impact of computers to understand how inevitable the nature of this change is. But as computers have largely been the dominant technology of the 1970s it is arguable that robots and automation will be the dominant technology of the 1980s.

Many people would argue that there is nothing new in automation—the Victorians were masters of mechanical automation. However, whereas traditional automation has been dedicated to one job or process, the 'new automation' is reprogrammable and is therefore adaptable to many jobs or processes. Robots belong to this new automation and this book is written to provide practical guidance for those who are considering introducing robots into their working environment.

As yet robots are a relatively new phenomenon, and although they have been commercially available for over twenty years, they have been traditionally regarded with some trepidation as many early robots did little to inspire confidence in the observer. Their movements were jerky, they were inaccurate, difficult to program and their reliability left much to be desired. But as with much technology, robots could only get better.

Improved mechanisms, control circuitry and computing power have all led to the point where nearly all robots are basically sound units that can be used with confidence. These improvements are reflected by the increasing use of industrial robots worldwide. The explosive growth in recent years has largely taken place in Japan and in the USA, but it is the success of the Japanese that has encouraged Western industrialists to follow suit and invest in the new automation.

Robots are used in an incredible variety of applications. There are few manufacturing processes where a robot could not be used, although whether or not it is practical and economic to use one is another argument. They appear in the most unlikely places: in the sea, in space, in coal mines and inside nuclear reactors. However, this book is largely aimed at the more mundane aspects of robot usage in manufacturing industries. In this context it is worthwhile examining what robots are actually used for today.

Rather than simply listing the applications, to facilitate comprehension it is useful to classify robot applications into:

(a) the robot as a tool user,
(b) the robot as a work handler, and
(c) the robot used for assembly.

In the first classification of robot usage the robot most closely emulates man. The tool to be used is mounted on the end of the robot arm and the robot is taught the most efficient program possible to complete the work. The robot then reproduces the program on request. Robot applications in this classification include: flame cutting, grinding, spot welding, arc welding, spraying, finishing, sealing, inspection/testing, marking, and glueing. In all of these applications a robot is more productive than a man, and in some (flame cutting, grinding, spot welding, arc welding, spraying, glueing) there are strong environmental and safety reasons for using a robot.

Overall there is a high element of 'skill capture' in these applications, and with skilful operations the man reacts to many and subtle prompts from the working environment and from the work process. Because of this, great care has to be taken the ensure the work is presented to the robot correctly and accurately. In addition, much work has been done on developing sensing devices for robots so that they can become more adaptable to their surroundings and workpiece variations.

In the second classification of robot usage the robot actually handles the work rather than the tool. This classification falls into three natural subdivisions:

- The robot manipulates work (forging, fettling, investment casting). This range of work includes some very unpleasant working conditions and heavy manual labour, and are thus good indications for the use of robots rather than people.
- The robot loads or unloads the workpiece to complete a process (casting, pressure diecasting, hot and cold pressing, injection moulding, heat treatment, glass cutting, soldering and brazing, machine loading).
- The robot moves the workpiece between workstations (conveyor-to-conveyor transfer, palletising, stacking, packing, sorting).

In the latter two categories of robot application many of the boring, arduous types of work are found, where the fatigue factors on human operators tends to limit the output of manufacturing technology.

It is arguable that the third classification of robot usage (i.e. assembly) ought not to exist separately, simply because so much of assembly work is encompassed in other classifications. Within a workcycle an assembly robot may: manipulate a tool, transfer work, inspect and test, and assemble components. But because one robot does all of these, and more, is argument in itself for a separate classification.

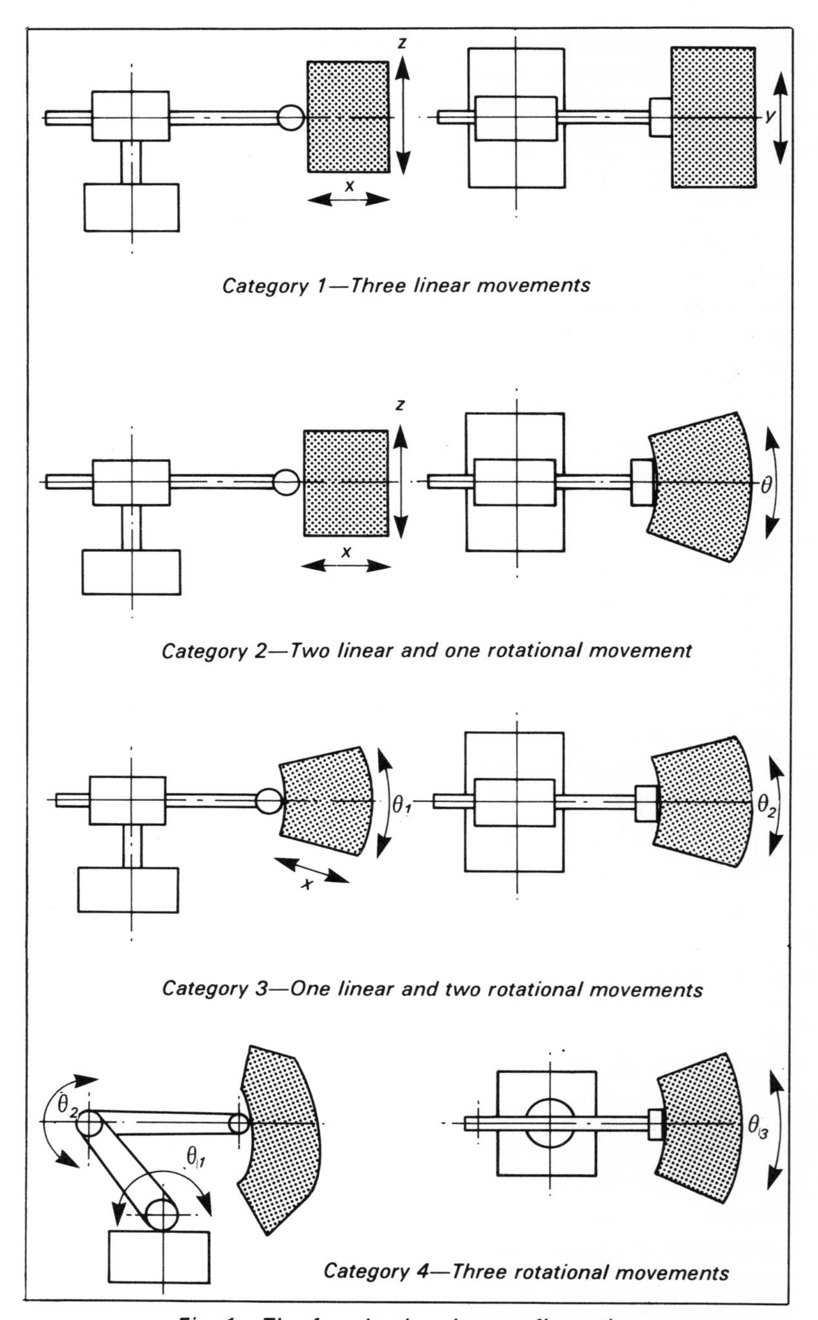


*Fig. 1 The four basic robot configurations*

It is clear that the potential for robot assembly is vast and nearly every industry has at least one assembly operation. Why then are there not more robots doing this kind of work? The answer to this question lies in the human operator's ability to compensate for bad fits and tolerances, for awkward component presentation and for unexpected events.

Naturally much work is being done to enable robots to overcome these limitations. Engineers and designers are now beginning to design for robot production rather than manual production and, as has already been mentioned, sensing devices are being continually developed to give the robot true interaction with its environment.

Considering all of these classifications it is clear that there are still many unexplored areas of potential robot usage that have not been identified. Every day adds a new page to the book of robot applications and it is this which makes robots so interesting, and which should encourage all manufacturers actively to consider if they need a robot.

Clearly no single robot would be suitable for all of the applications that have been mentioned. In fact there are four basic robot configurations, defined in terms of the three basic axes and the particular combination of linear and rotational movements used (Fig. 1). Nearly all robots have two or three additional axes that are mounted sequentially on to the basic axes. With the size and power of these various combinations the working volume and payload of the robot is defined.

The first three of these configurations were earliest to be developed for commercial use. Category 1 is particularly suited to applications that require precise linear movements, such as those required for assembly operations. The limitation of this category is that the $y$ axis is difficult to extend to any great length without considerable problems in supplying power for the motors that are used. This problem was overcome to a greater extend when the $y$ axis was converted into a rotational axis.

Thus in category 2 and category 3, $\theta$ and $\theta_2$, respectively, considerably extend the scope of the robot within quite a limited floor area. This makes these categories very popular in work-handling applications.

Category 4 types of robot are relatively recent arrivals on the scene. They require a high level of mechanical design and sophisticated control algorithms. However, size for size they have a correspondingly larger working volume, and with greater articulation have better access to awkward jobs. These features make these robots particularly suited for tool-handling applications.

As new technical innovations become available they are usually added to existing ranges of robots, and features such as linear and circular interpolation are now quite common. Undoubtedly robot vision and touch sensing will be equally common in the not too distant future. From a manufacturing point of view it is almost certain that robots will become as popular as computers are at present.

The problem with 'crystal-ball gazing' is that it does not solve any of today's problems. The assertion has already been made that this book is written for people who are new to robot technology and who wish for some practical guidance to use robots. With this in mind the book has been structured as follows:

*Part I—Defining the need for a robot system and chosing the correct robot.*

This part of the book takes the reader through the various stages of analysis, prescription and prognosis associated with robot system purchase.

*Part II–Areas that require special attention.*

These chapters are designed to enlarge on some of the more difficult subjects of safety, personnel matters and the financial appraisal associated with the first three chapters.

*Part III—Case studies.*

The final part of this book is dedicated to five illustrations of robot applications from all over Europe. They have been chosen to demonstrate a variety of interesting applications. Each case study takes a structured walk through the application, bringing out relevant details of interest to the potential robot user.

Throughout this book emphasis has been placed upon the use of tried and tested techniques that have been used effectively by the author and by people, who like the author, have earned their living by designing and installing robot systems. To the sophisticated reader many of these techniques will seem basic. Nonetheless they are still worth considering even if in some circumstances they are replaced by some other techniques that are more suited to an application or industry. Nothing in this world is guaranteed, but if the advice offered in this book is intelligently applied then the probability of successfully applying a robot is greatly enhanced:

"A wise man learns from others' mistakes—a fool from his own"

# *Part I*

*Chapter One*

# Defining the need for a robot

A ROBOT system, if it is to be effectively used, must meet a real need in a company. In the cold financial light-of-day a robot system should be a 'need-to-have' not a 'nice-to-have' piece of equipment. Without the focus of the need-to-have environment there is not the incentive to make the robot system work effectively and efficiently. Too often in the nice-to-have situation robot systems do not make an effective contribution to production or profits and end up as objects of curiosity in a corner of the factory.

There are many ways of defining the need for a robot system. Some people wait until they are forced into using robots because of health or safety problems. Some react to a crisis and buy a robot system in the belief that it will effect a solution. Others decide in a fairly arbitrary way to use a robot for a task simply because they feel that it is a good thing to do. In many instances these approaches work and therefore cannot be totally discounted.

However, it is the characteristic of good management to plan their activities within a structured framework. Within this philosophy lies the approach of beginning with corporate objectives and through a process of rational analysis ending up with a well-defined need for a robot. This approach is conventionally called a 'top-down' approach, whereas the previously mentioned techniques tend to be reactive or 'bottom-up' approaches.

Taking a top-down approach imposes a natural structure on this chapter. The first part discusses how corporate objectives can be translated to measure the potential for robot system use.

The second part considers a business from a robot point of view to discover some of the problems and risks the robot system could present. These two sections lead to a natural decision point where the company can decide to undertake a feasibility study or abandon the project altogether. (Section 1.3 is an abridged example of such a study.)

## 1.1 Corporate objectives and robot systems

Every organisation, whether large or small, has objectives. Sometimes they are expressed formally and sometimes they are simply the personal philosophy of an entrepreneur. It is these objectives that give a company direction, that give workers something to identify with and give managers ways of measuring the company's performance.

By and large business objectives are developed in response to the surrounding business climate (whether national or international) and economic climate. They are statements of corporate response that should be framed to give a formula for survival and growth. Strategic objectives tend to be framed in relatively abstract terms that are related to one aspect of the company's activities. The following are hypothetical examples that might be stated:

*Financial objectives*

— the company must achieve 40% return on any capital invested.
— the company must increase its turnover from £2 million to £6 million within five years.

*Market objectives*

— the company must increase its share of the soft-goods market from 0.5 to 3.5% within two years.
— the company must develop three new products for the fast-food market over the next 18 months.

*Production objectives*

— there must be an overall drop in scrap components from 15 to 5% per annum during the next financial year.
— the company must reduce its direct labour costs by 15% over the next year and still maintain existing production levels.

The problem with these statements are that they say *what* is required without saying *how* it is to be achieved. (It is the role of the manager to formulate a reply to the latter question.)

In the context of this chapter the particular task is to assess how a robot system can help to achieve corporate objectives. So, if a robot is to be justified at all, then it must be justified in terms of the objectives. Taking the previously listed objectives, the justification must be termed as follows:

- The capital investment required for the robot system is £100 000. Discount cash-flow calculations (see Chapter Six) show that this system will generate a return on investment (ROI) of £53 000 over the next five years. This represents 53% ROI and therefore exceeds the corporate minimum of 40% ROI by 13%.
- The use of robots in the final assembly area and in the spot-welding sections will relieve the existing production bottlenecks. It has been calculated that the effect of this investment will be to increase throughput by 200%. Additional investment will also be required in the forming department. Two further punching machines at a total cost of £250 000 are required. The cumulative effect of this total investment will be to triple the output of the factory during the next three years.
- Analysis has shown that the major reason that the company has difficulty in achieving penetration of the soft-goods market is its inability to react to the rapidly changing consumer demands. As a result ways of building much greater flexibility into the existing production system must be found. With this additional flexibility it will be possible to increase market share substantially, certainly beyond the 3.5% corporate target. It is therefore recommended that the whole production system be analysed to see where robots may be applied.

These examples show how the corporate objectives can be translated into performance criteria for a robot system. Similar statements can be derived for the other objectives.

It is only in relation to these company objectives that the performance of a robot system should be specified and measured. To get from the company objective to the final statement that details the way in which the robot will achieve the target, requires quite a detailed analysis, commonly called the feasibility study (see Section 1.3).

This is inevitably quite a time-consuming activity that can be very expensive. Because of this it is as well to be reasonably certain that the whole exercise is going to be worthwhile. Fortunately this can be done simply and quickly by considering how well the characteristics of the company match those of robot systems.

## 1.2 Company and robot system characteristics

The point has already been made that robots can be used in a great variety of applications, but as to whether they are the best solution must be assessed in different circumstances and individual situations. It is rarely a simple decision and what is required is to examine the way in which the company functions to see how well it matches the innate characteristics of robot systems.

Clearly there are many possible company characteristics that could be chosen for the purposes of analysis. In practice it turns out that there are relatively few that are sensitive to robot usage and these may be summarised as follows:

(a) the way in which the company markets affect product life cycles,
(b) the type of production methods the company uses, and
(c) the organisational climate.

What must be examined is the way in which these characteristics react with the basic robot characteristics of:

*Flexibility*—the ability of the robot system to cope with many jobs within one workcycle, and to change between one workcycle and another rapidly.

*Consistency*—the ability of the robot system to consistently repeat workcycles.

*Speed*—the ability of robot systems to perform some aspects of work more quickly than conventional manual techniques.

*Environmental indifference*—the ability of robot systems to work in a wide variety of noxious, arduous and dangerous environments with indifference.

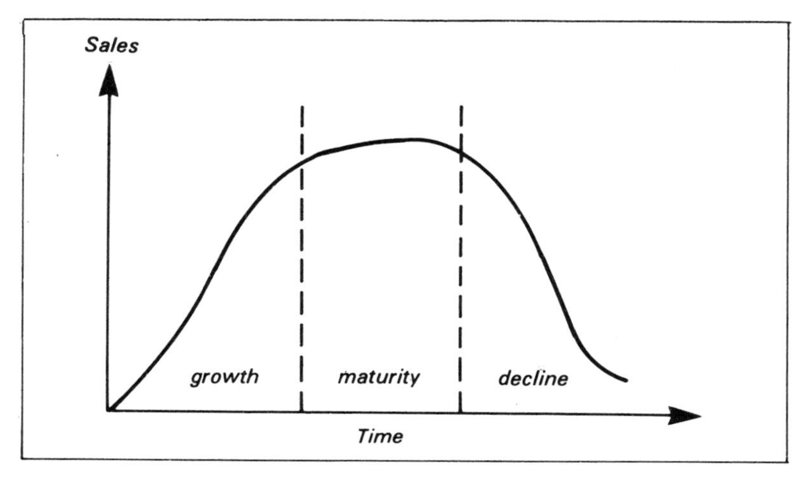


*Fig. 1.1 Product life cycle*

**The market place and product life cycles.** The sole purpose of any production facility must be to produce the right goods at right place at the right time. In this way the company can sell its products and make a profit. Unfortunately the market place is a notoriously volatile environment and there are many influences at work that affect the sales of a company's products.

Some products are subject to seasonal fluctuations in demand (e.g. umbrellas and woollen hats), some products have natural yearly cycles (e.g. fashion products), and some products are subject to technological change (e.g. micro-electronics). Whatever the reason and however long or short the life of the product, it can to some extent be measured and predicted.

Typically the life cycle of a product goes through three distinct phases: growth, maturity and decline (Fig. 1.1). The length of these cycles can vary considerably depending upon the products that are being sold.

The majority of companies do not have only one product but have many products which are usually at different parts of their life cycles (see Fig. 1.2). This is necessary in order to even out demand on finite productive resources and to stabilise the total income of the company. Because of the continually changing demand for these products there are constant

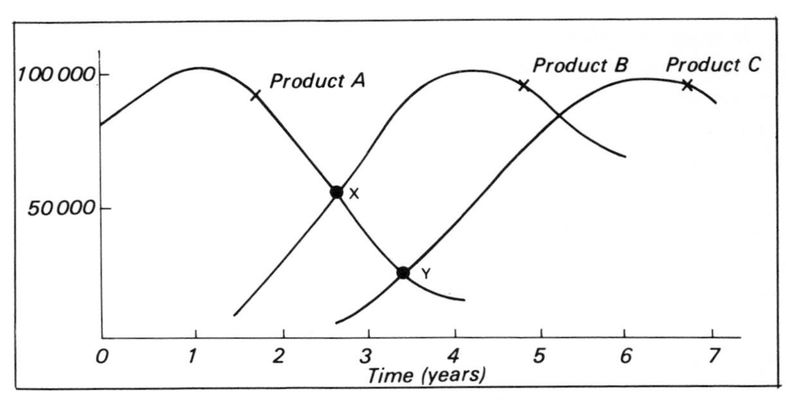


*Fig. 1.2 Life-cycle overlaps*

problems of allocating productive resources. At point X on the diagram there is a demand for 60 000 units of product A and 60 000 of product B. At point Y there is a demand for 15 000 units of A, 15 000 units of C and 90 000 units of B. Taken as a whole there is a fairly constant demand of about 120 000 units from the production system.

Where there are dedicated resources for a product there are few allocation problems; however, where there are shared productive resources there are endless problems of production balancing that often require sophisticated computer programs. Feeding new products into an existing system compounds these problems. New tools, jigs and fixtures have to be designed and made; new drawings made and new work methods devised and introduced. If these problems are to be overcome then there have to be ways of reducing the inertia of the conventional manufacturing system.

Recently the use of numerical control systems has greatly assisted in the area of machine tools and machine centres. Such systems allow production to be efficiently run using the batch techniques that are necessary in the multiproduct environment. However, these solutions can only ever be a part of the total drive to reduce manufacturing inertia. Other equipment and techniques are available to further shorten the total manufacturing system response time. Computer-aided design (CAD) can shorten the design phase and computer-aided manufacture (CAM) can help to integrate and optimise system utilisation, but these aids are inevitably limited by the flexibility of the system.

It is robots that can provide much of the flexibility that will be required to maximise total system output. They are able to move work around a factory and between workstations. They can either replace operators or make them more productive. In short, they provide a sort of 'manufacturing oil.'

With this increased responsiveness a company is in a position to react to market pressures and requirements very quickly, thus assisting in business planning, market penetration and exploitation.

**Production types.** The problems of using robots in very small or very large batch manufacturing systems are worth greater consideration. Generally, if batch sizes are very small and nonrepetitive (i.e. 1–100), then the setting-up time and the teaching time associated with using robots may impose too heavy an overhead. Conversely if the batch sizes are very large (say 100 000–500 000) and repetitive, then fixed automation ought to be considered. An important modifier to the batch size is related to the volume of work that is done by the robot and the number of times a batch is repeated.

Consider, for instance, the manufacture of bulldozers. The number of bulldozers made in any year is quite small—probably no more than three to seven per week from any given supplier. Clearly the batch size is small but if the amount of arc welding that is done on some of the large fabrications or the bulldozer is considered, it can run into 10–20 man-hours per fabrication. Robot productivity in the arc-welding environment is about three times that of a man's productivity; therefore using a robot could reduce the cycle time for arc welding to 4–6 hours.

Providing that the robot can be fully utilised welding a number of other fabrications then it will make sense to use the robot even on small batches. In a similar way to the manufacturer of bulldozers, there are many industries which only manufacture small batches of products. However, because these batches are repeated every day, week or month over a reasonable period of time there is an opportunity to amortise the programming and setting-up overheads for the robots over a large number of batches. In these circumstances it can also be viable to use robots.

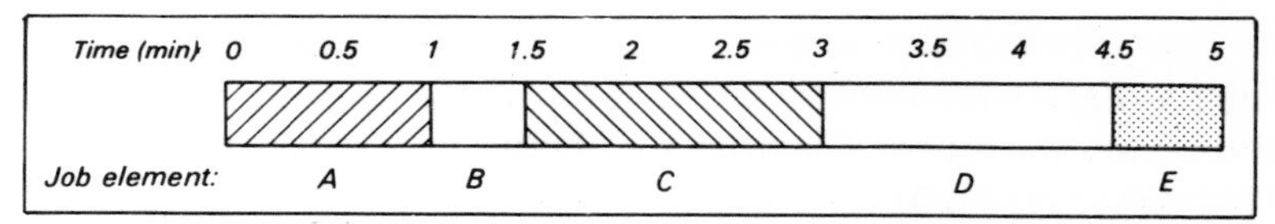


*Fig. 1.3 Example of a low-efficiency production cycle: A, load; B, close safety guard; C, robot positions itself (transition time); D, robot performs task; E, unload*

On examining different types of batch production it becomes evident that the key to using robots is the workcycle itself. In all installations maximum robot-system utilisation should be aimed for. This is achieved by keeping the robot working as much of the time as possible. Some workcycles have innately low efficiency characteristics in terms of the relationship between the operation to be done by the robot and the non-productive work that has to be done before and after the operation. Sometimes limits are set by the surrounding machinery or processes (Fig. 1.3). Using the example in the figure, efficiency (expressed as a percentage) is calculated from:

$$\frac{D}{A + B + C + E} \times 100 = 43\%$$

With this kind of work characteristic it is clear that in order to efficiently utilise the robot it would be necessary for the robot to service two or three workstations. If the application mitigates against this course of action then it will be necessary to carefully examine the economic case to be made for using the robot, as it is almost certain to be very weak.

As job element D in Fig. 1.3 gets larger in relation to the other elements the efficiency of the workcycle increases and the justification for the robot is less likely to be dependent upon a multi-workstation concept. The converse of the situation occurs when the robot is used to load and unload a workstation. The longer it takes to perform these tasks the greater the benefit that will be gained from using a robot, as the robot's speed will help to reduce the overall manufacturing cycle time.

Taking an overall view it is clear that along with the batch sizes that are to be processed, and the repetition of these batches, the nature of the workcycle is equally important. The total combination gives a very strong indication of the nature of the robot system and the nature of the savings that can be made by using the robot system. With this information, management is in a position to assess the economic benefits that are associated with the robot system.

**The organisation and change.** There are very few organisations that can accept robot technologies without some degree of internal stress. For some unknown reason robots seem to induce a higher level of stress than other more conventional types of automation. This stress has the effect of working against the production benefits that can be gained. Since at this stage commitment to robots is limited then it is as well to examine:

(a) the company's technological profile,
(b) the skills of the existing management, and
(c) the state of development of the company's management control systems.

If a company has experience of handling modern technologies then it is likely that robot systems can be assimilated without fuss. The skills that are required to maintain, adapt and innovate with robot technology already exists within the company. If a company is not a high-technology company then the process of education and familiarisation is inevitably quite extensive. In addition there is also likely to be the need to buy-in relevant technical skills in order to harness and maximise the benefits of using robots.

Until control of this nature is established (and it is best established before any equipment is bought), then the level of doubt or uncertainty among operators and managers often leads to the situation where the slightest problem with the robot system leads to a crisis.

The necessity of operator training for new technology is obvious and is discussed in Chapter Five. However, the training of managers is frequently overlooked simply because managers do tend to be self-motivating people who can look after themselves. As far as new technology is concerned it is best to train managers in the same way as operators to ensure that their level of understanding is adequate and enables them to operate efficiently. Certainly the traditional skills of management will not be redundant, but additional skills will be needed to cope with a new type of man–machine interaction. With these factors in mind a useful exercise is to have a personnel audit undertaken to anticipate the type and magnitude of likely problems.

The final issue to be raised is that of management control systems. It must be remembered that the majority of existing control systems have been designed with manual production in mind. Built into these systems are many assumptions about the way in which the production system as a whole functions. In other words these systems have a 'model' that may not be effective when robots are used. Generally this becomes evident when unexpected bottlenecks occur in departments either before or after those in which the robots are to be applied. This implies that it may be necessary to revise these control systems before the robot system is bought. Alternatively, if no management control systems exist then they ought to be introduced before the robots are introduced. Also, it is good practice to ensure that any system whether existing or to be purchased is flexible enough to be adapted to cope with the flexibility of robot production.

The purpose of this section has been to introduce an air of reality into anticipating potential problems that may mitigate against the use of robot systems. Can the company adapt to new technologies? Will the organisation's control systems adapt to the use of robots? These are important questions which will in turn prompt other questions—all of which must be faced. Some of the problems are generated because of the nature of the business—which cannot be changed. If this is the case then can a robot truly meet an organisational target? Sometimes circumstances can be changed or manipulated to make the use of a robot favourable, then it could be right to go ahead. If the answers to the questions are all positive then it is also advisable to undertake a feasibility study.

## 1.3 Feasibility study

The purpose of a feasibility study is to undertake a structured analysis of a production system to determine if, where and how a robot system (or systems) will help achieve a company's objectives. The scope, time spent and cost of a feasibility study will normally be in direct relationship to the quality of the results. By and large it is worthwhile putting a lot of effort into the feasibility study as it can considerably reduce the elements of uncertainty connected with the decision to buy a robot system.

The best practical approach to a feasibility study is to complete it in three distinct phases:

(a) the analysis,
(b) evaluating the alternatives, and
(c) choosing one course of action.

The analysis is a top-down exercise. It starts with a summary of the whole production system. Then the individual process to be considered for robotisation is examined to determine where robots should be best used. This phase leads naturally to the evaluation of possible solutions to the problem. To keep an air of reality and perspective both robot and non-robot solutions should be included in this phase. Finally, one solution should be chosen that best suits the company's objectives. It should include the relevant details upon which a decision can be made.

This whole exercise can be structured as follows:

(i) A statement of the company's objectives.
(ii) An outline study of the overall production system, including the products made, the organisation of the factory, and the areas to be included in the feasibility study.
(iii) A detailed study of the robot application or applications being considered. This should include details of the organisation of the relevant areas, the methods employed at present (who does what, when, where and why?), a description of the control systems, and the working systems used at present.
(iv) A summary of the feasibility, advantages and disadvantages of using a robot in each application.
(v) Investigation and discussion of available robot systems.
(vi) A brief summary of alternative methods that have been considered and why they are not suitable.
(vii) A detailed description of the changes that are required to the existing manufacturing system that will be necessary to introduce robots. This should include the identification of any product redesign where it is necessary to facilitate the use of robot systems and also spell out any implications for the company's marketing policies.

(viii) A description of the recommended robot system together with details of the costs associated with purchase, installation and running the system. An implementation plan that identifies the major activities and their estimated time should be included in this section.
(ix) A financial appraisal of the proposed system with appropriate financial calculations.
(x) A summary of the way in which the proposed robot system will achieve the company's objectives.

The work involved in this chapter takes the potential robot user to the point at which a system can be specified, ordered and installed. There should be no doubt in the buyer's mind about the reason for using a robot system and the benefits expected from it. Along the way to these conclusions will have been an opportunity to examine the total production system. This examination in itself will be a useful exercise in which a fresh perspective of the company can be gained. From this perspective, opportunities other than robot systems may emerge, and perhaps the company's overall objectives can be considered in the light of new technologies. Old engineering designing systems and techniques that are orientated towards manual means of production can be reappraised and if necessary changed. Old control systems can be overhauled and new opportunities within the company and market place can be explored.

# *Chapter Two*
# Choosing a robot

THE PROCESS of choosing a robot for an application is not easy. If the wrong robot is chosen then it can be an expensive mistake, not only in terms of cost but also in terms of wasted time and effort. The reasons that people experience difficulty in selecting robots are as many and varied as potential applications; however, more often than not they fall into the following categories:

(a) uncertainty about the benefits expected from using a robot in a particular application,
(b) unfamiliarity with robot technologies and the way in which technical data sheets express the information required for an application,
(c) difficulty in translating the information that is available about robots so that the choice can be narrowed down in a meaningful way, and
(d) uncertainty about the requirements of a particular job or process.

If the case for using a robot has not been clearly established then it is arguable that it is pointless proceeding any further. If, however, an analysis (such as that described in Chapter One) has proved the need for a robot then the way forward is clear and the problems associated with points (b), (c) and (d) above can be addressed.

A straightforward logical approach that can be undertaken in a series of simple steps may be summarised as follows:

*Stage 1*

| manual application description | + | improvements objectives | = | robot application description |
|---|---|---|---|---|

*Stage 2*

| | | | | |
|---|---|---|---|---|
| robot application description | + | technical translation | = | robot search description |

*Stage 3*

| | | | | |
|---|---|---|---|---|
| robot search description | vs | many robot specifications | = | few suitable robots |

*Stage 4*

| | | | | |
|---|---|---|---|---|
| few suitable robots | vs | testing and applications trials | = | robot for application |

## 2.1 Stage 1—Preparing a robot application description

The work undertaken during the preparation of the feasibility study is all that is required to complete this first stage. The feasibility study will examine existing work processes, and then against company objectives describes what will be required of any equipment that is to be used for a particular application.

## 2.2 Stage 2—Preparing a robot search description

The purpose of this stage is to convert the information gathered during Stage 1 into a form that can be used to select a robot. This is clearly a nontrivial task, bearing in mind the wide variety of potential robot applications and the wide variety of available robots. Fortunately robot specifications usually all contain the following type of information:

Number of axes.
Type of movement of each axis (e.g. linear, etc.).
Speed of movement of each axis (expressed in $m\,s^{-1}$ or degrees $s^{-1}$).
Range of movement of each axis.
Description of the working volume.
Cumulative accuracy of the robot (expressed as ± 0.X mm).
Repeatability of the robot (± 0.X mm).
Gross weight of the system.
Load capacity (quoted either as being applied at the end of the last axis, or as acting at a point beyond the end of the last axis).

Motional power source.
Type of memory (e.g. volatile or nonvolatile) of the controller.
Size of the memory of the controller (e.g. 32 K) and whether the memory can be extended.
Power requirements of the robot system.
Maximum operating temperature of the system.
Maximum operating humidity of the system.
Number of interface channels available.
The type of interface channels (e.g. voltage drivers, current drivers, relays).
Method of teaching (e.g. manual, lead-through, point-to-point).
Safety features.
Process synchronisation.
Software adaptability.
Special packages (e.g. spot-welding packages, rewelding packages).
Special features (e.g. extra axes, system integration facilities, etc.).

To many, these features will seem confusing, but in reality they are not. So before proceeding further it is worthwhile enlarging on each feature in order to more fully understand what information is required to create an 'ideal' robot profile.

**Number of axes.** In practice, most robots have between three and six axes (five being the average). Not all applications require five or six axes; for instance, work transfer may only require three axes, whereas paint spraying may require five or six.

**Type of movement of each axis.** There are only two basic types of movement, linear and circular, and these exist in various combinations. There have been other articulations devised, most notably the 'elephant's trunk.' These are usually developed to a particular application, which in the case of the elephant's trunk was paint spraying. On balance there is little to suggest that these developments are significantly better than, say, adding another axis.

**Speed of movement of each axis.** This is always expressed in either degrees $s^{-1}$ or m $s^{-1}$, depending upon whether movements are radial or linear. It is often necessary to translate degrees $s^{-1}$ into a linear measure in order to relate it to a process requirement. Where many axes move simultaneously it is difficult to calculate the precise transit time between two points. The problem here is to account for the time it takes for the acceleration and deceleration of the manipulator within a particular motion. In practice if times are this critical they will be measured manually. Sometimes robot software contains linear and circular interpolation facilities. (Linear interpolation is when a robot describes a straight line with the wrist end of the manipulator between two points within its working volume; circular interpolation is when a robot describes a circle through three points defined within its working volume.) It is normal for the precise speed of movement for these interpolations to be defined when the robot is programmed.

**Range of movement of each axis.** This is quite simply how far an axis will move either in degrees (circular axis motion) or in metres (with linear axis movements).

**Working volume of the robot.** Since robots have many axes, the working volumes often have many convolutions. This is particularly true of robots with all radial movements. Since the work to be completed on an application has to reside within this working volume it must be accurately described.

**Accuracy and repeatability of a robot.** The accuracy and repeatability of a robot are always measured at a predetermined point at or near the end of the last robot axis. (Sometimes the accuracy will be quoted at the end of the tool that a robot holds.) The accuracy of a robot relates to the robots ability to return to a taught position after programming, measured at the end of the last axis, within its working volume. Repeatability is a measure of the robots consistency and accuracy over a period of time. These terms are often interchanged but repeatability is always a better indication of the stability of the robot system.

**Weight of the system.** This is of importance where floor loading is critical.

**Load capacity of the system.** This is a practical limit placed by the design of the robot mechanism and the control circuits. Often it is expressed by manufacturers in terms of moment, e.g. 10 kg acting 50 cm from the tip of the last axis. Sometimes it is quoted as a total load capacity (including the weight of the gripper mechanism) measured at the gripper–arm interface. Exceeding the capacity usually causes a downgrade of the accuracy and repeatability figures.

**Motional power source.** There are three basic types of robot motivational sources: pneumatic, hydraulic and electric. By and large robots tend to only have one type of motivational source. The main reason for this is an engineering one, in that it is simpler to design an all-electric machine, an all-hydraulic machine or an all-pneumatic machine. Each motional power type has particular advantages and disadvantages—pneumatics are cheap, but have poor power-to-weight characteristics; electrics are expensive, but have good speed response and accuracy characteristics with average power-to-weight ratios. Hydraulic robots have good power-to-weight characteristics and are relatively cheap, but they often have only moderate accuracy and repeatability characteristics.

**Type of memory in the controller.** There are two types of memory associated with the robot control computers: volatile and nonvolatile. Volatile memory is memory which is erased when the robot system is powered down or when an emergency power-cut occurs. Nonvolatile memory is not erased on power down. All robot systems have both volatile and nonvolatile memory, but have different functions. The nonvolatile memory keeps all of the information about how the system works and what jobs have been taught. The volatile memory is where calculations are done and the system is controlled.

What often confuses people is when the phrases RAM, ROM and PROM are quoted in brochures. The active memory of a computer is volatile and is described by so many K of RAM (random access memory). The long-term memory is nonvolatile and is described by so many K of ROM (read only memory) or PROM (programmable read only memory). When 'K' is quoted it relates to the size of the memory.

**The size of the memory and whether it can be extended.** The size of of the memory has a direct relationship to the performance of a robot system. Again the difference between volatile and non-volatile memory is important. The larger the volatile memory, the more sophisticated the software can be. The more sophisticated the software, the better the control and the more extensive the features such as the linear interpolation can be. The larger the nonvolatile memory the greater the number of jobs that can be held. Often nonvolatile memory can be extended by the use of disk storage or magnetic-tape storage. In this case, the size of the fixed nonvolatile memory is only important in that it limits the size of the program that can be taught.

**Power requirements of the robot system.** This is usually expressed in kilovolt amps and relates to the power used by the robot. Where the robot is used with other equipment this must be determined separately.

**Maximum operating temperature of the system.** Determines the temperature at which a robot's performance begins to seriously deteriorate.

**Maximum operating humidity of the system.** Determines the level of humidity at which the electrics and mechanics of a system begin to malfunction.

**Number of interface channels.** No robot works in isolation and the number of interface channels, both input and output, can greatly limit the flexibility of the system.

**Type of interface channels.** Determines what interfaces can be made directly and indirectly. For integration with processes and other parts of the productions' system, voltage and current drivers and receivers are necessary. Where there has to be a physical decoupling of the system, relay contacts are necessary. Where digital information has to be passed between computers or from sensors external to the robot system, then standard digital interfaces (either serial or parallel) exist.

**Method of teaching the robot.** Given the extent to which the success of a robot depends upon how well it is taught, the method of teaching is very important. Most robots when taught are put into a teach mode and either taught manually or on a point-to-point system. Typically, manual teach systems are used for paint spraying. It is often called 'teach-by-doing' and can be quick and effective, providing that a high level of accuracy is not required.

Other applications requiring higher levels of accuracy are usually taught by the 'point-to-point' system. In this system the robot is under power when taught. A teaching pendant is connected to the robot controller and each axis is controlled by a separate potentiometer. Then by adjusting the potentiometer, the operator takes the working end of the robot through the route required, stopping every few centimetres to plot a point. With linear interpolation, providing the route required by the robot is straight, the points plotted can be up to 200 or 300 cm apart. This is a great saving in both programming time required by the operator and memory required in the computer memory.

**Safety features.** The laws which relate to the safety of manufacturing equipment quite clearly expect equipment to be fail-safe. This is particularly important during the teaching phase of using robots as man and machine work in close proximity. Any emergency should leave the robot in a state that presents no threat to anyone standing nearby. Also, it should be possible to interface safety interlocks (sensing devices or emergency stop buttons) that enable the robot system to be put into a safe state in the event of unexpected occurrences. It is as well to make sure that the safety requirements are understood and that they are adequate for the installation as a whole, and the application and environment.

**Process synchronisation.** Many robots are used in applications where the work is moving on a conveyor or transfer system. Where this is the case and the robot is required to coordinate its movements with the passing work, it automatically transposes the taught program. Process synchronisation is, in a way, a part of software adaptability.

**Software adaptability.** As a general concept software adaptability is usually recognised by the ease with which programming changes can be made. As a feature it is important in batch work where continuous change occurs.

| *Robot specification* | *Robot application description* | | | | | | | | | | | | | | | | | | | | |
|---|---|---|---|---|---|---|---|---|---|---|---|---|---|---|---|---|---|---|---|---|---|
| | *Physical layout* | *Work movement macro* | *Work movement micro* | *Manual constraints* | *Work variety* | *Work sequence* | *Work loads* | *Space available* | *Services available* | *Safety required* | *Process requirements* | *In-work interfaces manual* | *In-work interfaces machine* | *In-work interfaces control* | *Between station interrupt manual* | *Between station interrupt machine* | *Between station interrupt control* | *Servicing skills* | *Training required* | *Manual skills* | *Floor loading* |
| *Number of axes* | | ● | ● | | ● | | | | | | ● | | | | | | | | | | |
| *Axes type* | ● | ● | ● | | ● | ● | | | | | ● | | | | | | | | | | |
| *Axes speed* | ● | ● | ● | ● | ● | ● | | | | | ● | | | | | | | | | | |
| *Axes movement range* | ● | ● | ● | ● | ● | ● | | | | ● | ● | | | | | | | | | | |
| *Work area* | ● | ● | ● | | | | ● | ● | | | ● | | | | | | | | | | |
| *Accuracy* | | ● | ● | | ● | | ● | | | | ● | | | | | | | | | | |
| *Repeatability* | | | ● | | ● | ● | ● | | | | ● | | | | | | | | | | |
| *Weight limit* | | | | | | | | | | | ● | | | | | | | | | | ● |
| *Load capacity* | | | | | ● | | ● | | | | ● | | | | | | | | | | |
| *Movement type* | | ● | ● | | ● | ● | ● | | | ● | ● | | | | | | | | | | |
| *Memory type* | | | | | | | | | | | ● | | | | | | | | | | |
| *Memory size* | | ● | ● | | ● | ● | | | | | ● | ● | ● | ● | ● | ● | ● | | | | |
| *Memory extension* | | | | | ● | | | | | | ● | | | | | | | | | | |
| *Power type* | | | | | | | ● | | ● | ● | ● | | | | | | | | | | |
| *Operating temperature* | ● | | | | | | | | ● | | ● | | | | | | | | | | |
| *Operating humidity* | ● | | | | | | | | ● | | ● | | | | | | | | | | |
| *Number of interfaces* | | | | | ● | ● | | | | | ● | ● | ● | | ● | ● | | | | | |
| *Interface type* | | | | | | | | | | | ● | ● | ● | ● | ● | ● | ● | | | | |
| *Teaching method* | | | | | ● | ● | | | | | ● | | | | | | | | | | |
| *Safety factors* | ● | ● | ● | ● | ● | ● | ● | ● | ● | ● | ● | ● | ● | ● | ● | ● | ● | ● | ● | ● | ● |
| *Process synchronisation* | | ● | ● | | | ● | | | | | ● | | | | | | | | | | |
| *Software adaptability* | | | | | ● | ● | | | | | ● | | | | | | | | | | |
| *Packages* | | | | | | | | | | | ● | | | | | | | | | | |
| *Special features* | | | | | | | | | | | ● | | | | | | | | | | |

*Fig. 2.1 Preparation of robot search description*

**Special packages.** Many robot manufacturers have links with manufacturers of other equipment and produce fully integrated packages. This saves the prospective buyer having to worry about buying from different equipment manufacturers and then having to integrate a system themselves. In most cases these packages are successful and the only serious drawback is the possibility of having little choice in the type and manufacture of the ancillary equipment offered.
**Special features.** As with all commercial products, robots have special features that become a part of product differentiation. They rarely affect the fundamental operation of the system, but represent the 'nice-to-have' rather than the 'need-to-have' aspects.

Given that these categories have direct relevance to the choice of a robot, then the way to complete each category description is to relate it to the information that has been gathered for the robot application description. By analogy, the buyer is rather like someone who has gone into the library looking for information about a series of subjects. The suitable procedure is to look under the book cataloguing indexes, locate where they are in the book shelves and then extract the information. The same procedure can be adopted when choosing a robot: the books on the shelves are represented by the robot application description, the questions to be asked are the categories of robot description. What is required is the index and map.

Such an index and map is shown in Fig. 2.1. If accuracy is taken as an example, the corresponding robot application description categories are: work movement macro, work movement micro, work variety, work loads, and process requirements. Interpreting this in practical terms:

- Work movement macro defines the limits of the large movements required from the robot. So, if a robot has to move work between two conveyors two metres apart then it must have an operational range in excess of two metres.
- Work movement micro usually relates to the movements or dexterity required of the robot within a particular task. These can be demanding (as in the case of assembly tasks) or simple (as in the case of lifting a sack).
- Work variety dictates the range of accuracy required between jobs.

- Work loads dictate the range of loads to be carried and this will limit the expected accuracy. For example, a robot that moves cement packs will not require the same accuracy as one that has to insert integrated circuits.
- Process requirements are constraints that are not specifically related to loads or load carrying capacity of the robot. Instead they are imposed by the manufacturing process itself, e.g. spot-welding will require the weld to be placed within 1–1.5 mm from its predetermined position.

Clearly this process is simple providing the information is available. In terms of the library analogy, the books have to be on the shelf before the information can be extracted.

On a broader plane, two factors emerge from Fig. 2.1: safety is something which affects every aspect of the work profile, and the work process requirements affect every aspect of the robot profile. In simpler terms, safety needs to be designed into the system right from the start, and the work determines the choice of robot—not vice versa.

## 2.3 Stage 3 and Stage 4—Narrowing down the choice of robot; testing and application trials

With Stage 2 completed the process of narrowing down the choice of robot is now virtually a mechanical operation. The 'ideal' robot profile is simply matched against specifications provided by robot manufacturers and those which have the best fit can be short-listed. Ideally that would be the end of the search, however in reality it is often necessary to apply an additional test. This is usually in the form of a simulation of the work to be completed or an actual sample of the work.

If the test is to be a simulation of the work then it is important that the simulation includes the most critical aspect of the work. For example, if the robot is to be used for assembling integrated circuits on to a printed circuit board, then the most critical aspects of this work will be actually putting the 'legs' of the integrated circuit into the holes of the printed circuit board. Conceptually this might be represented by inserting a row of 8 pins with a diameter of 0.5 mm into a row of holes 0.6 mm in diameter.

To convert this into a robot test simply get a robot to pick up suitably mounted pins, move them a distance that they might typically move in a real working environment, then insert the pin assembly into predrilled holes in a board of the same shape and size as the real printed circuit board. To be more realistic the test should be extended by drilling more holes over the test board and inserting the pins at differing positions.

If the test takes the form of undertaking an actual sample of the work, it will usually be done by the robot supplier on their own premises. Enough samples should be provided for a convincing number of repetitions to be completed. Also, suitable jigs and fixtures should be provided for reasonably accurate simulation of the real working situation. This form of testing is often chosen where the robot is actually wielding a tool (as a man would). Arc welding and paint spraying are good examples of this. In both cases the choice of sample should reflect the complexity of the usual work passing through a production unit.

Much can be learned from tests of this type. In both cases, if the teaching process is done in front of the potential user then some idea of programming techniques and difficulties can be gained. Enough time and samples should be available to enable the robot manufacturers to prepare a demonstration to the best of their ability. There is no sense in rushing this evaluation as the next activity will be to buy the robot if it performs adequately. Where problems occur with either simulation or real work demonstrations, approach them with an open mind. It may well be that some product redesign or process redesign will be required. Failure is not always the fault of the robot and valuable clues to the nature of the change required can often be gained.

Within the logical structure described in this chapter, a robot may be chosen, and the user can be reasonably certain that mistakes will be avoided. To summarise:

(a) From the description of a manual system define what is expected from a robot system.
(b) From this expectation prepare a profile for a robot.
(c) Select a few suitable robots.
(d) From testing and application trials choose one robot to do the task.

As this process naturally follows on from the activities described in Chapter One, completing it will take the user through a 'one-pass' operation, thus avoiding costly delays to reinvestigate some aspect of production previously omitted.

# *Chapter Three*
# Specifying, ordering and installing the robot system

THE LOGICAL progression of the exercises in Chapter One and Two lead to the point at which the necessary basic analyses have been performed. Upon these sound foundations it is possible to proceed to the subsequent stages of specifying, ordering and installing the robot system with confidence.

There should be little doubt as to how the robot will dovetail into the manufacturing process. The impact of the robot on the remainder of the manufacturing system will be understood and where necessary programmes of remedial action can be planned and implemented. Performance targets for the whole system will have been quantified.

The remaining tasks are to negotiate satisfactory agreements with trade unions and the workforce (see Chapter Five), and design adequate safety measures (see Chapter Four). Once these are completed, or at least underway, the system specification can be undertaken.

## 3.1 System specification

The purpose of the system specification is to:

(a) describe in precise terms what the performance of the individual items in the system has to be,
(b) describe how these individual items will interface and operate as a system, and
(c) to put (a) and (b) together in order to cost the system construction and to invite tenders from equipment and system suppliers.

The first part of this process will in part already be completed in that the specification for the robot will have been done (see Chapter Two). In order to complete the robot specification, the process and ancillary equipment performance will have to be known and defined. As these are established the remaining point to be considered before part (a) is complete is that of future requirements. This affects the specification in that it will dictate the upper and lower limits of the system's performance. The source of information for this exercise will be the company's marketing plan.

As the company grows or diversifies, so strains will be put on productive resources. Since robots tend to be used to achieve expansion or increase productivity, then they must impose as little constraint on growth as possible. Conversely it would be folly to overspecify the performance of the equipment. For if the equipment specification covers all eventualities then it will probably be far too expensive. Thus a balance must be sought that is cost effective initially and at the same time allows sufficient room for growth.

Marketing plans will give a good indication of the rate of growth of the company and this can be related to the robot system. So, for example, if a product is processed at a workstation in three minutes and the marketing plan projects a tripling in output, then the robot system must be capable of completing the process in one minute and, if used, conveying equipment must move the work at three times the original speed.

Sometimes, as is the case when robots are used for tool handling (e.g. arc welding or spot welding), then the target of tripling output cannot be achieved simply by making a robot work three times faster than a man. This is because the constraints imposed by the process will dictate that even though the robot may move three times more quickly than a man the process parameters required to match this speed cannot be created. In these cases the intention of the marketing target can only be achieved by using two robots. However, often these need not necessarily be purchased simultaneously and capital expenditure can be phased in line with actual growth achievements.

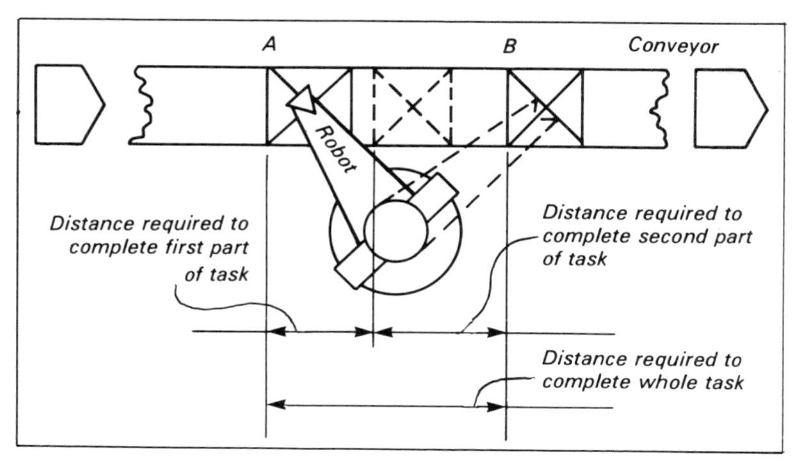


*Fig. 3.1 A robot working on a production line assembly task. If a robot dictates the assembly process cycle then the slowest part of the robot workcycle will dictate the upper limit of the operating speed of the work system*

What emerges from this kind of exercise is the need to categorise, in a simple way, the role a robot plays in production. If it is to be used as a part of a production line then the upper and lower performance limits will be dictated by the production line itself. The actual speed of the production line will be dependent upon the largest or slowest part of the assembly or process cycle. If a robot dictates this cycle then the slowest part of the robot workcycle will dictate the upper limit of the operating speed of the work system.

This is simply illustrated by a robot working on a production line assembly task (Fig. 3.1). The distance from A to B in the figure is related to the time it takes the robot to perform both of its tasks operating at its maximum speed. The track cannot, therefore, move any more quickly otherwise the work will not be completed. If this unit were to be used for paint spraying and not assembly, then the time it takes the work to move between A and B will be dictated by the process, i.e. the paint can only be delivered at the speed that gives the correct finish without running or missing areas of application. In this instance the rate of paint application will be dictated by the type of paint used, its viscosity and the pressure of the air used to disperse the paint.

When robots are used in situations where they are decoupled from other production units, then the performance characteristics of the robot tend to be defined by the production or by the nature of the tool being used by the robot. In this way they are similar to the previously quoted example of paint spraying.

Clearly, no matter what the robot application, this kind of simple classification can greatly assist the setting of performance characteristics. What has now to be described is the nature and range of the interfaces between the pieces of equipment that form the robot system.

**Defining the interfaces.** As with perforance considerations the definition of interfaces is greatly simplified by the classification of robot applications and relating these classifications to the previous concepts of the robot in the production process. The classification of robot applications has already been done in the introductory chapter, that is, (a) the robot as a tool user and (b) the robot as a work handler. The manufacturing status, is of course, (a) the robot used in the production line and (b) the robot used in a decoupled situation.

By relating these combinations to real-life applications the categorisation may be completed as is illustrated in Appendix A. An examination of this appendix illustrates the range and possibilities of interfaces that can connect seemingly simple pieces of equipment. The choice of interface will be decided by technical issues rather than managerial decisions.

There are different reasons for choosing different interfaces. Digital signals are popular because they will talk directly with computers that control systems of this type. As a general rule serial digital interfaces are more popular than parallel digital interfaces because they offer higher noise immunity, and therefore greater reliability. This noise immunity can be very important in systems that are associated with volatile processes like arc welding. Analogue electrical signals are commonly used to control such variables as speed and temperature. These signals also have good noise immunity characteristics.

Optical interfaces are usually used in situations where there is a necessity to avoid physical contact between pieces of equipment or even between parts of the same piece of equipment. Optical devices are usually to be found associated with speed measurements or angular displacement measurement. It is usual to find that for transmission purposes these signals are converted into digital signals.

Finally, mechanical interfaces are often specified when one piece of equipment has to interlock with another. This category of interface takes many and varied forms, ranging from simple electrically operated bolts to microswitches or limit switches.

Whatever type is used (in any circumstances) will obviously be the choice of the system designer who will always use 'reliability' as the governing criterion.

**Specification example.** So far much of the discussion has been of a general nature exploring the underlying reasons for issues that affect system architecture. To bring these considerations into focus the specification of an overhead conveyor for a paint-spraying system is considered. An example of a small system is shown in Fig. 3.2.

From this diagram it can be seen that the painting process has a number of different phases: loading (A–B), robot 1

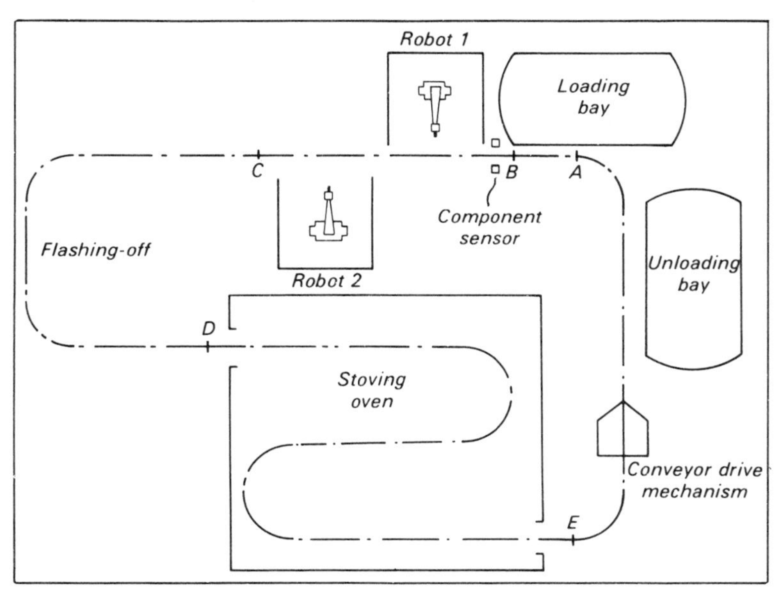


*Fig. 3.2 Example of an overhead conveyor for a paint-spraying mechanism*

spraying side 1 of the component and robot 2 spraying side 2 of the component (B–C), flashing-off, i.e. solvent evaporation (C–D), and stoving (D–E). Each of these phases will require a finite amount of time to complete and will be a function of track length and track speed.

Considering the specification of track length, it can be seen that the distance from A to B has to be long enough to allow the loader space and time to load-up components and perhaps build up a small buffer. B to C will be dictated by the size of the robot spray booths. C to D will be determined by the length of time it normally takes the paint solvents to evaporate. Sometimes this process is aided by passing the components through a warm air unit and these are often used where space is limited. D to E has to be long enough for the stoving process to be completed for the range of components to be fed on the system. Finally, the distance from E to A must allow for the components to be unloaded and inspected. Typically, for a small system, A to B will be 4 m, B to C 6 m, C to D 40 m, D to E 40 m and E to A 10 m, giving a total conveyor length of 100 m.

The speed of the track will be related to the time required for the flashing-off and stoving processes. These will be determined in turn by the type of paint used and by the size of the components to be processed. For large components the track speed must be low, perhaps as low as 0.521.0 m min$^{-1}$. For small components the speed will be higher, perhaps as high as 1.5–2.0 m min$^{-1}$. Therefore, allowing for some lattitude a practical working range could be 0–2.5 m min$^{-1}$.

Whether or not this range has to be continuously variable can be decided by the range of product variation. A widely varying load will require continuous variation, whereas a few standard lines will probably only require the speed to be varied in steps of 0.25 m min$^{-1}$.

Clearly as speed is closely related to the size of the components so will the arrangements for hanging the components from the conveyor. Conventionally components are hung from either wire hooks of varying lengths or from hangers. Sometimes special frameworks are used so that the component density is kept as high as possible. Whichever is the case the spacing of the mounting is another variable that must be specified.

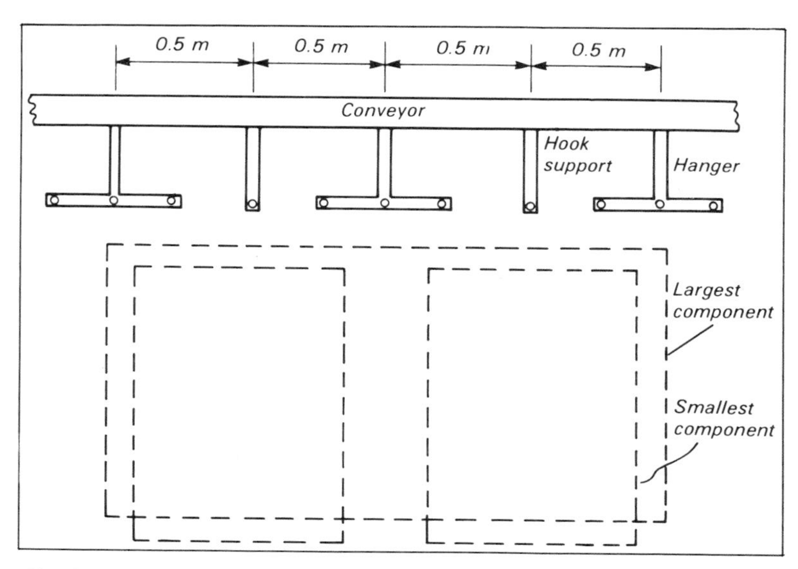


*Fig. 3.3 Possible working arrangement for hanging components on a a conveyor system. For a 100 m track it will be necessary to specify 100 hangers and 99 hook supports alternately spaced on 0.5 m pitch*

The calculation of this spacing is a simple affair. The largest and the smallest discretely hung components that will have to be processed need to be determined. If for the sake of this example it is assumed that the largest component is 1.95 m long, 1.0 m deep and 0.5 m wide, and that the smallest component is 0.75 m long, 1.0 m deep and 0.15 m wide, then it is clear that hook mountings will need to have an extreme pitch of at least 2.0 m. Within this extreme, smaller components will also have to be hung in ways that maximise packing density. Since it is possible to hang two 0.75 m components within a 20 m span then it seems that an obvious subdivision would be to have hook or hanger spacing at 1.0 m. Additional flexibility will be gained by further subdivisions of the 1.0 m space. (Fig. 3.3 illustrates a possible working arrangement.)

Closely associated with the way in which the components are hung on the conveyor will be the load carrying capacity of the conveyor. In calculating this it will be necessary to ensure that the maximum weight of components is taken into consideration. From the example given in Fig. 3.3, it can be seen that the maximum loading will occur when 100 of the smaller components are hung on the conveyor (i.e. 100 × 40 kg) rather than when 50 of the larger components are hung on the conveyor (i.e. 50 × 60 kg).

This weight will not be a single-point loading but evenly distributed around the track. To calculate the load that the driving motor will have to cope with the weight of the hangers, the weight of the chain and the effective weight of friction will have to be added together. In total this figure is likely to be as high as 8000–10 000 kg. The distributed static load will have to

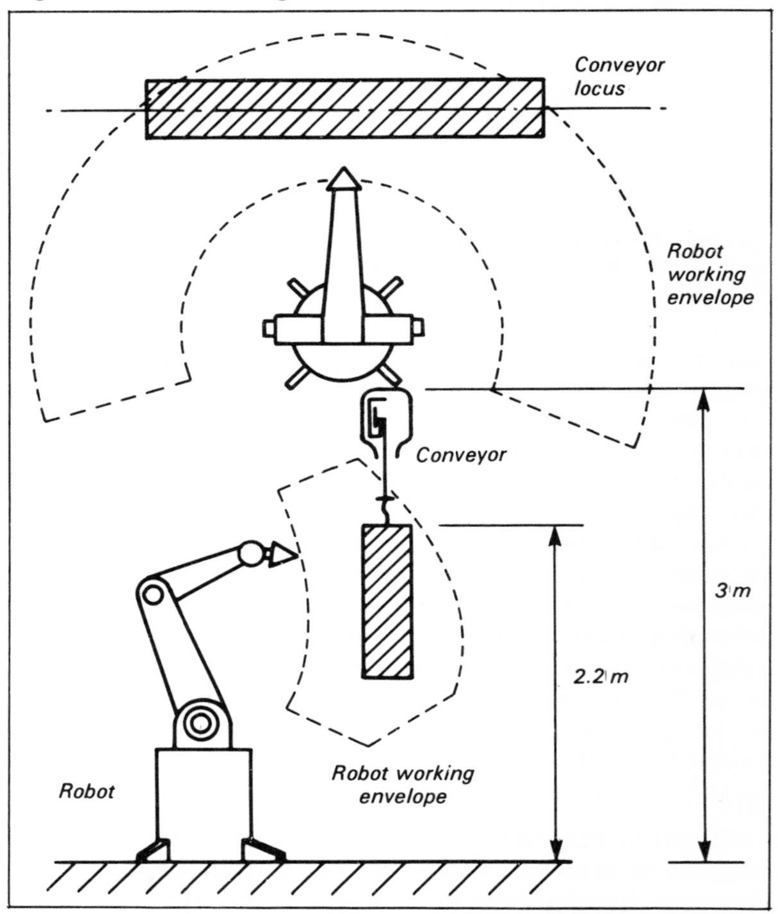


*Fig. 3.4 The height of the conveyor from the ground will be determined by the working envelope of the robot*

be calculated to ensure that the maximum floor loading is not exceeded at any of the track support beams.

The height of the conveyor from the ground will be determined by the working envelope of the robot. Clearly the largest component has to be mounted so that, as far as possible, it stays within the robot working envelope. This is illustrated in Fig. 3.4. It can be seen from this diagram that if the component is mounted in the position shown, the top of the component will be about 2.2 m above the ground. Allowing 0.6 m for the mounting devices the track will have to be mounted 3.0 m above the ground.

The question of defining the interfaces that will be required to integrate the conveyor into the remainder of the paint-spraying system still remains. The interfaces are best considered in two parts—those that are required to pass information to the remainder of the system and those required to receive information from the system. The outgoing interfaces will need to pass information about the status of the conveyor, its relative position with respect to the robots and a safety interlock. The incoming interfaces are likely to have to cope with speed change requests, safety interrupts and external stop/start requests.

Typically the conveyor status will be a simple switch which when closed indicates that the conveyor is functioning correctly, and when open indicates a malfunction. The relative position of the conveyor can be generated in a number of ways. The diagram of the layout of the system shows some component sensors. These are usually optical devices that are triggered from the hangers or from the components. The signal that is generated from this device is usually transmitted as a digital signal. The outgoing safety interlock will normally be required to conform with the safety standards of the system.

In the majority of systems of this nature the normal requirement is for a closed electrical contact that opens and latches when activated. The input interfaces usually have to respond to requests or demands from the remainder of the system. It is usual to signal a change in speed by varying a voltage to generate an analogue signal. Therefore the conveyor controller will need a signal line that can translate this signal into action. From a specification point of view the only information required is to specify the range of voltage variation. Commonly this voltage range is 0–10 V.

Safety interrupts are the reverse of the outgoing safety arrangement; that is, if it sees that the safety link is broken it triggers the correct shut-down procedure on the conveyor. This signal will nearly always be a closed contact. External stop/start are usually electrical contacts that will trigger automatic circuit makers or breakers.

To enable a competent supplier prepare a quote the conveyor specification example can be summarised as follows:

| | |
|---|---|
| Conveyor length | 100 m |
| Conveyor height | 3 m |
| Conveyor speed range | 0–2.5 m $min^{-1}$ |
| Distributed maximum load | 4000 kg |
| Maximum hanger load | 60 kg |
| No. of hangers | 100 } alternately spaced |
| No. of wire hook supports | 99 } alternately spaced |
| Hanger–hook support spacing | 0.5 m |
| Power supply | 440 V (3 phase) |

It cannot be denied that putting robot systems together is at times a complicated task requiring some experience and considerable expertise. Because of this, and because a company's own expertise is sometimes inexperienced with automation, many choose to subcontract subsequent phases of the system completion to external agents. When this is the case then it is usual to commission a 'turnkey' system from either the robot supplier or from a firm of system engineers.

A third option which is often overlooked is to use a firm of project engineers/managers who will supply experienced engineers and managers to work alongside the company's own engineers and together build the system. This approach has the singular advantage in that it grafts experience into the company which is not subsequently lost when the project engineers depart. This of course will pay dividends in the future when system modifications are required or when a new robot system has to be designed.

The reason for raising this issue is connected with the way in which the system specification is translated into an order. By now there is enough information available to price each component of the system. With this the base cost of the system can be calculated; that is, the cost without special interfaces or computer software or of system engineering. All these items add cost to a system. If the company's own engineers have been used then an accurate estimate of the man-days required to complete these tasks can be made. This of course can be costed and used as a measure against which turnkey quotes can be judged.

## 3.2 Ordering the system

At this stage most companies would choose two or three suppliers to submit quotes for either the complete system or for the individual system components. Once received, the company is in the position to judge which quote represents the best value for money. As a word of caution it is worthwhile asking those suppliers who submit tenders for complete systems for convincing evidence of previous experience. This can in part be provided in the way of previous installations and partly by satisfactory evidence of the depth and quality of the engineering and management experience they can devote to the project. Unless this evidence is quite convincing it is probably unwise to take a chance on the lure of a cheap quote.

Once the equipment supplier or system supplier has been chosen the remaining task is to place the orders. Again it is wise to counsel caution. All orders must have some contingency built in to ensure that the company's completion date, budget costs and technical performance requirements are achievable. Paring away these contingencies can often eliminate the ability to manoevre when things go wrong in the implementation phase of the project. In particular, caution must be taken in the area of cost. Usually this has a direct relationship with quality and reliability. Robot systems are usually designed to last 10 to 20 years and a compromise on quality and reliability should not be readily accepted.

With all of the purchasing activities it is as well to bear in mind that quality and reliability should always take priority over cost.

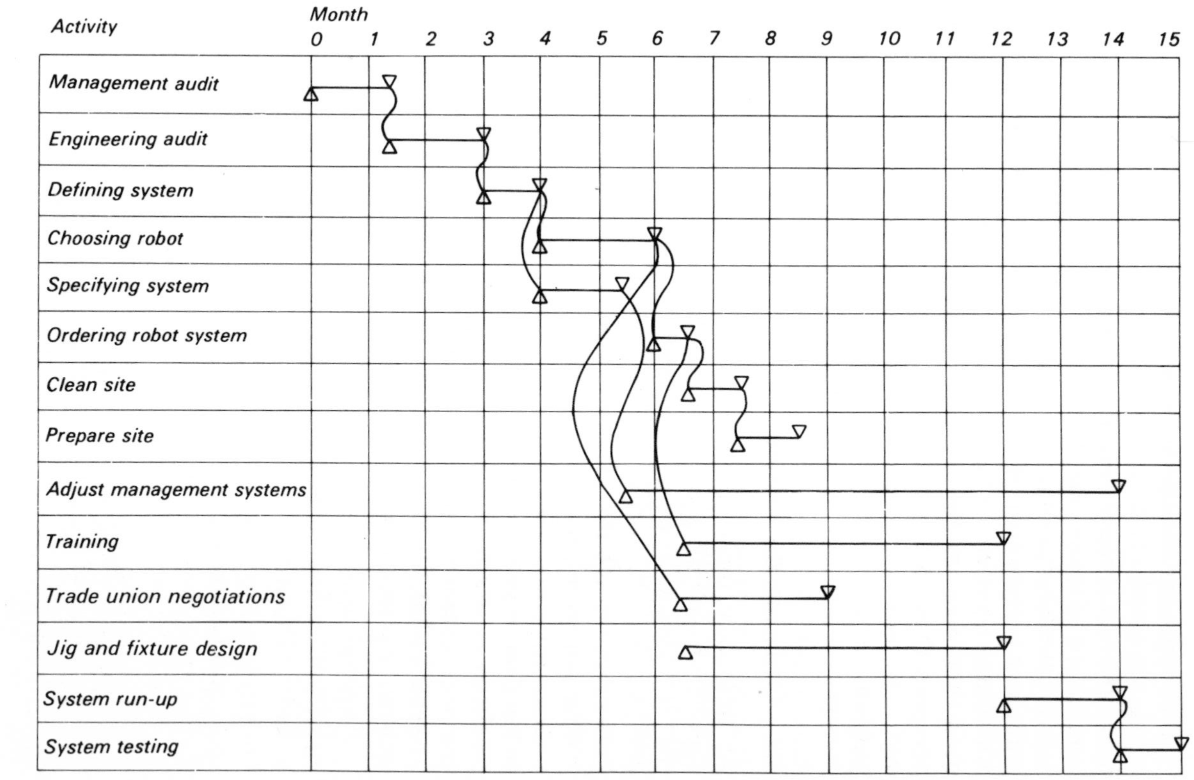


*Fig. 3.5 Example of a system-implementation bar chart*

## 3.3 Preparing for the system

Once the tenders have been received and the choice of supplier (or suppliers) made, then work must begin to prepare for the robot system's arrival. The starting point for all system implementations is the preparation of a work schedule or plan. Of course many companies would have completed a plan right from the start of the whole process; this activity is highly recommended.

However at this stage a plan becomes a necessity not a luxury, because of the large number of parallel activities that have to take place. The plan must take the company through the preparation for the system to the running-up of the system. It is possible to identify the major activities involved during these processes and a standard technique such as critical path analysis or bar charting (see Fig. 3.5) can be applied. The work usually subdivides into the following activities:

Breaking down the existing system.
Preparing the robot site.
Rearranging the existing production facilities.
Revising production control and planning systems to maximise the robot utility and achieve the targets established for the system.
Creating buffer stocks and rearranging storage facilities.
Training/retraining/redeployment/redundancy.
Negotiating trade union agreements.
Designing and building new jigs and fixtures.
Reviewing existing quality control techniques.

These activities will not necessarily take place in this exact sequence but each is important and deserves a more detailed examination.

**Clearing the site.** Since few robot systems are installed on 'green-field' site the usual starting point for most systems is to clear the proposed site. It is surprising how often delays occur because someone has forgotten to move some existing equipment or dismantle an existing facility. The larger the robot system the less likely this is to be a problem simply because the installer usually has to 'shoehorn' the new system into a restricted area.

Thus extra care should be taken to remove any possible obstacles. In smaller systems odd items tend to be forgotten. Overhead pipework is left to interfere with the new system, power rails and trunking can be trapped by new rails and trunking. Sometimes obvious items such as ventilation ducting are overlooked. All of these problems are simply overcome by meticulously inspecting the proposed site to ensure that nothing has been forgotten.

**Preparing the site.** Once the clearing exercise has been completed then work can begin to prepare the site for the new robot system. By and large these activities should be organised on a ground-up basis. So work begins by ensuring that the floor is flat and if necessary is reinforced to support the equipment that has to be installed. Then the necessary services, such as air, electricity, ventilation, gas and drainage are installed.

Only now is the equipment installed, but even at this stage it is not always the robot that takes precedence. Often when a robot is being used to load or unload a workstation, such as is the case when handling forgings, it will be one of the last items to go in. But whether first or last, the whole process has to be carefully organised to ensure that there are no oversights that could cause delays.

**Rearranging existing production facilities.** With all of this activity going on it is not unnatural to find that existing production processes become disrupted. More often than not this tends to occur when the robot system is directly replacing an existing manual system. When this is likely to be the case then it is best to try to arrange for existing production to be temporarily moved elsewhere on the existing site. When this is not possible, and this is quite often the case, then it may be necessary at some time to temporarily transfer the manufacturing facility to another site. Alternatively, as a last resort, the new system may have to run-up in other premises and then transferred.

**Revising production control and planning systems.** Once the decision has been made to go ahead and purchase the robot system then this should be the trigger for a major review of the existing supervisory management systems. In particular the production control and work planning departments need special attention. The reason for selecting these two activities outlies in the need to: (a) keep the robot supplied with work throughout the working day (which is often extended to two or three shift workings in order to minimise the pay-back period), and (b) use the robot as efficiently as possible when it is working by ensuring high utility.

It is surprising how much strain these requirements can put on systems that have been evolved to cope with manual production. Quite often the robot system tends to trigger the purchase of some more sophisticated computerised systems that can match the increased inertia that the robot system introduces into the manufacturing cycle.

**Changes in jigging and fixturing.** Alongside the revision of supervisory systems will be inevitable changes in the techniques of jigging and fixturing. It is paradoxical that while offering such flexibility in potential application, robots require a much higher standard of design ability from the engineering staff in a company. This is largely due to the fact that manual system shortcomings can be accommodated by people using their common sense when they perceive a problem. Often people need only report a problem to the engineers after the event has occurred. Unfortunately robot systems are not blessed with 'common sense' and any intelligence by way of judgement has to be built into the system.

This particular problem becomes very evident when designing jigs and fixtures. Unlike a man a robot requires direct access to the work in order to do it efficiently. Thus it often becomes necessary to find new methods of clamping and work presentation, or even new sequences of manufacture in order to compensate for the robots apparent lack of flexibility. The great compensation for this increase in discipline is that jig and fixture designers become more skilled and there is often a corresponding pay-off in the design of normal manual methods of manufacture for nonrobot work.

**Reviewing existing quality control techniques.** By the same token the changes that take place in the area of manufacturing techniques and methods need to be mirrored in the disciplines of quality control. As the robot speeds up workflow so the possibility of producing scrap at speed increases. Changes will be required in inspection techniques that have to be carefully evaluated. Where possible it is best to design some quality checks within the robot workcycle. These should be integrated so that the errors are flagged to an operator and/or the robot workcycle is stopped.

The other personnel issues previously mentioned in relation to training, etc. require a closer examination and are dealt with in Chapter Five.

## 3.4 Installing the system

The fruits of all of the analysis, hours of planning and preparation ripen when the components of the robot system are delivered for installation. Assuming that these arrive according to a planned sequence, then adequate time will have been allowed for each system component to be tested against specification. There is no point in accepting substandard performance from any equipment as this only stores up problems for the future. In making this point it is necessary to recognise that in the process of installation timescales tend to concertina towards the testing time. There is always the temptation to let something go in order to keep to schedule. Be pedantic and be sure:

- Do conveyors maintain their speed under load and can the speed be accurately adjusted?
- Do forges and ovens have the correct thermal performance?
- Do work handlers or robot grippers work correctly?
- Do the jigs and fixtures work correctly?
- Are the robots accurate and is this accuracy repeatable?

In short, be sure of the performance of any piece of equipment, no matter how small, before assembling the system. Then, as the equipment is assembled into the system, each interface has also be be scrutinised and its performance monitored. By following these arduous procedures it is possible to isolate and cure many problems that become difficult to diagnose once the system has been assembled. Never accept a supplier's word unless it has been proven to your satisfaction and in a repeatable and quantifiable way. Even if buying a system that has been bought as a turnkey system evidence of adequate testing must be sought.

**Testing performance.** Since the science of testing is well understood by the majority of engineers the work described above will pose few problems. However, when it comes to testing the performance of the robot there are few, if any, standard techniques that have been devised and which are universally accepted. The problems of checking accuracy and repeatability on equipment that performs in a three-dimensional space is nontrivial. In overcoming these problems when working with robots, the author has devised some simple, cost-effective techniques, which have an acceptable level of statistical validity.

When testing accuracy, whether or not a robot can return to a previously taught point needs to be determined. To test for repeatability, assess if the robot can continue to return to the same taught point over a long period of time with continuous operations.

Figure 3.6a illustrates a simple piece of equipment that can be fabricated in most workshops at little cost, or which can be bought commercially. This device should be firmly attached to an adjustable stand so that it can be moved around within the normal working volume of the robot, and a needle should be firmly attached to the end of the robot arm or gripper. With these two devices it is possible to resolve the robot accuracy and repeatability with respect to the $x$, $y$ and $z$ planes of operation.

In principle and practice the technique is straightforward and begins by covering each tube-end with cartridge paper (Fig. 3.6b). The robot is then programmed to approach each tube along its central axis. The object is to get the robot to just touch the centre of the pencilled cross without puncturing the paper. Once this program is completed the robot should be put on replay and the results observed as the robot does about 50 cycles. If the robot is 100% accurate and its repeatability is perfect then the needle will return to exactly the same point again and again. Any error or drifting can be clearly identified and quantified by careful observation. Careful note should be taken of both the results of this test and of the position of the equipment in the robot working envelope.

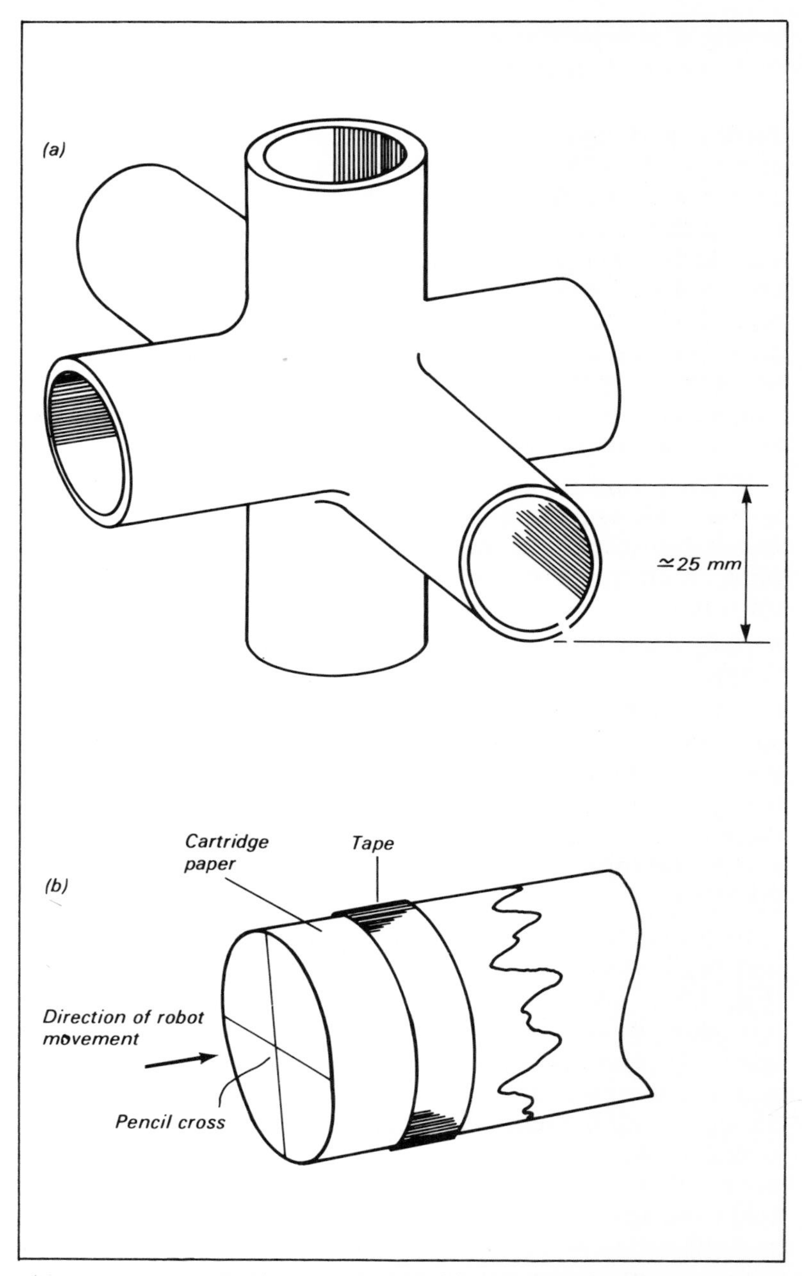


*Fig. 3.6 Device used for testing the performance of a robot's accuracy and repeatability (as devised by the author)*

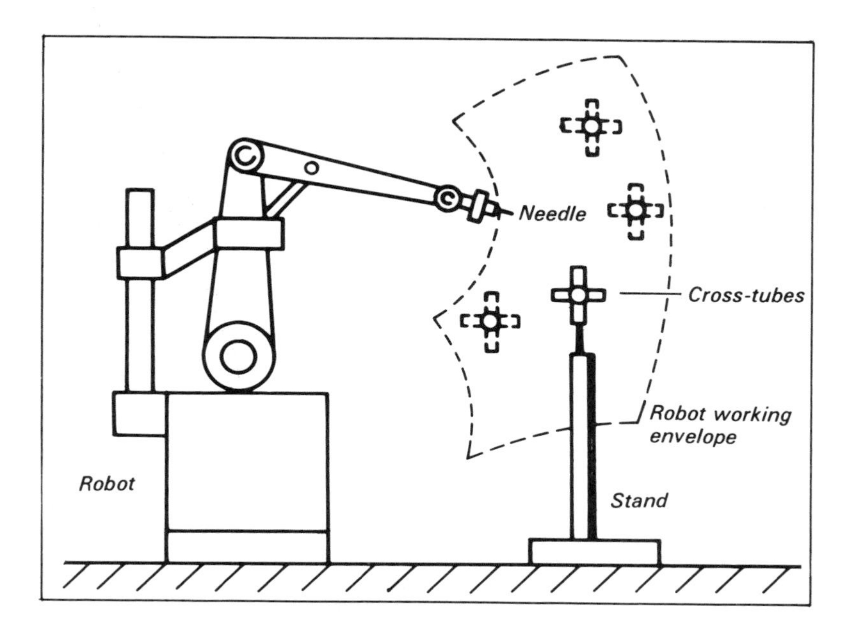


*Fig. 3.7 System for testing the performance of a robot's accuracy and repeatability. The cross-tubes are moved to another point within the working envelope and the process repeated.*

The cross-tubes should then be moved to another point within the working envelope and the process repeated (Fig. 3.7). A random sample of about 12 to 15 points, with the whole test being repeated at each point, will give a good indication of the overall accuracy and repeatability of the system. (Note, it is unlikely that any robot will be within specification throughout the whole of its working envelope and there is usually some deviation at the extremes of operation. With the results of these tests the majority of robot suppliers will be able to quickly rectify any faults that are found.)

Having 'pushed' the idea of components and system testing so hard it may seem to be labouring a point. It is not. If the system components work well then the foundation of the system is firm and the system can be run-up and commissioned.

## 3.5 Post-installation activities

In spite of the most stringent testing measures there are very few robot systems that work the first time they are switched on. The same is true of nearly every complex system. Components on printed circuit boards 'burn in' and sometimes malfunction, seals on hydraulic joints settle and begin to weep, loose wires are shaken from terminals, software errors are found in the computer programs, etc.

The cumulative effect of these errors and faults is a 'system settling-in time' which can be extremely frustrating. With correct testing this period should be minimised, even so it could last three or four weeks!

Compounding these problems are the errors that the operators will make simply because they are not familiar with the system (often referred to as 'finger trouble'). It seems that no matter how good the training that is given to operators there is always a learning curve related to the amount of 'hands on' experience the operators have. Fortunately the general experience of most users is that the problems associated with this learning curve, and those associated with the equipment running-in, tend to disappear almost simultaneously.

This time can be very depressing for both managers and operators. It is usually punctuated by a series of post mortems and arguments, claims and counter claims. However, in time the system and people settle down and morale is restored when the earlier expectations of the system are realised.

As the robot system increases its effectiveness, so the strain will be felt in other parts of the manufacturing system. At this point the work carried out to bolster the supervisory systems will begin to pay off. Other housekeeping activities will also have to be checked and where necessary bolstered. Maintenance schedules, fault reporting procedures, job-logging systems and correct safety and emergency procedures can be implemented and if necessary be modified in order to work smoothly.

This chapter may at times seem to have been rather pessimistic in outlook, especially when considering what may go wrong. It certainly could have been written to give the impression that all would go smoothly and with the minimum of bother. Indeed some systems actually do this. However, if something can go wrong it is the responsibility of the management and engineers to anticipate the worst and to plan for remedial action should the need arise. In order to plan for these possible events their nature and extent need to be known.

It is inevitable that there will be many learning curves associated with new equipment. Operators will have to become familiar with teaching programs and in new safety requirements. Engineers will have to learn what can and cannot be done with the robot. They will have to design or redesign products and associated processes to accommodate the changes. Managers will have to learn to manage technical resources of a more complex nature. However the optimism and excitement that is felt when the robot system functions well and productively, more than compensates for the problems that may occur. A move to use robot technology is a move towards the future and it is no surprise that the majority of companies who have the foresight to install one robot system usually continue to invest in other robot systems.

# *Part II*

*Chapter Four*

# Safety and robot systems

IT IS strange that so little has been written about the safety of robot systems. One or two articles appear from time to time about individual robot applications, but it seems to be a subject that both robot users and robot manufacturers are reluctant to discuss in detail. In practice, the majority of robot safety problems are resolved with common sense techniques that create both a safe and productive environment.

## 4.1 Safety—whose responsibility?

Irrespective of any legislation which may exist there is a clear responsibility on the shoulders of management to provide a safe working environment for their workforce. By the same token there is a reciprocal responsibility on employees to behave in a safe manner when at work. Accidents usually occur when either management or their workforce fail to honour the obligations they have to one another.

In the majority of the industrialised countries of the world there is legislation that formalises these moral responsibilities. This legislation usually describes in detail the safety precautions that should be observed generally in the working environment, and specifically for a particular machine or process. The problem with most new technologies, and robots are no exception, is that safety legislation does not keep up-to-date with progress.

The developer and applier of new technology can only be guided by common sense and general safety principles. So, safety precautions that are devised must be reasonably practicable and consistent with previous practices in similar technologies. Clearly it is too late to determine what is reasonably practicable after an accident. It is necessary to anticipate safety precautions and to design adequate safety features in every robot application.

The ability to anticipate the nature of the problems likely to be encountered, to a greater extent depends upon knowledge of the technology being used. As many new users of robots will not be familiar with robot technology there is an immediate problem. It is therefore a good starting point to briefly review the nature of robot technology in order to establish a common basis on which to begin the safety analysis.

## 4.2 The robot system—an overview

For the purpose of testing, maintenance and safety, robot systems may be regarded as consisting of a number of building blocks:

*The mechanical system*—is the most visible aspect from a casual observer's point of view. It consists of levers, pivots, beams and bearings.

*The motive system*—consists of motors and actuators that convert the raw power supplied from a prime power source (e.g. the electric mains) into motions of the mechanical system.

*The control system*—behaves rather like the human sensing system (i.e. smell, touch, sight, etc.), giving feedback into the computer system about where the robot is and what it is doing.

*The computer system*—is the processing element in the whole system. It has to coordinate the activities of the robot system to ensure that the system as a whole is responding in the desired manner.

*The software system*—coordinates the activity of the computer system behaving rather like a human brain.

*The interface system*—is the communications network that enables the robot to talk to the outside world and vice versa.

Collectively, these building blocks can function on their own. When they work together they become the robot system with a defined control priority—a control hierarchy (Fig. 4.1):

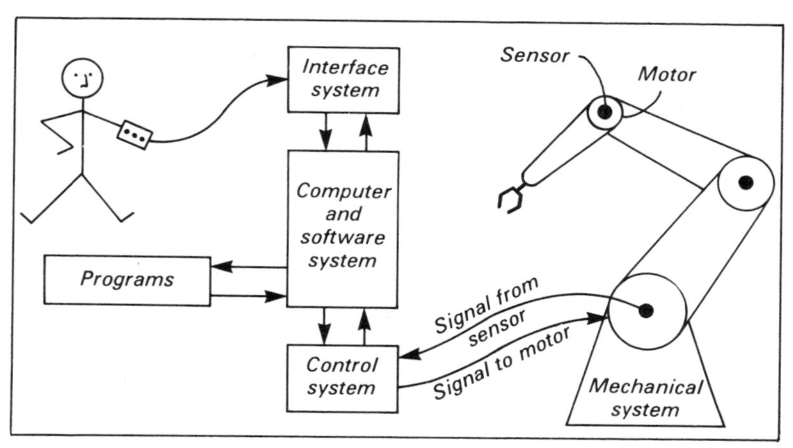


*Fig. 4.1 The control hierarchy of a robot system*

- The mechanical system is controlled and motivated by the motive system. The motive system translates hydraulic, electric or pneumatic energy (or any combination of these) into movements and hence work.
- The motive system is controlled by the control system which has a central element and a distributed element. The distributed element is located at pivots or bearings throughout the mechanical system.
- These elements are positional sensors and accurately translate positional information about the mechanical movements of the system. The sensors feed information into the central part of the control system (usually situated in the control console of a system). The central control system generates the order for a motor or actuator to move and then, via a positional sensor, senses whether the movement is correct.
- For the control system to generate its orders for the motivational system it needs to know what those orders are. These orders are provided by the computer system. The computer system examines the program of activities that have been entered by an operator. It translates this program into a form the central control system can understand. This translation is done via the software system.
- Finally, the program for the computer system is supplied via interface channels.

In a well-ordered system everything is in balance and the control hierarchy performs smoothly. However, because of the interdependance of the system elements a malfunction in one can cause the whole system to behave unpredictably. It is this unpredictability that can give rise to an unsafe situation. Understanding the nature and manifestation of this unpredictability enables adequate safety precautions to be designed. However, before this analysis is done it is necessary to fully understand the implications of the robot in the working environment.

## 4.3 Safety—the robot and the work process

It has already been established that robots are used in many different working positions and work processes. Each situation demands particular safety precautions for operators, equipment and the robot system. To illustrate these precautions consider the very common robot application of paint spraying.

As a process, paint spraying presents two distinct sets of problems. Those associated with health and those associated with the danger of fire. The paint spraying environment must be intrinsically safe. It is the health problems that make paint spraying a particularly attractive robot application. When men perform the task of paint spraying they must wear clothing that will protect them from the fine paint mists and wear ventilator masks that enable them to breath air free of paint molecules. The clothes they wear must be antistatic to prevent the possibility of sparks, and of course the smoking of cigarettes is prohibited. More generally the whole spraying environment will require forced ventilation to prevent the build-up of flammable vapours.

Simply replacing a man with a robot does not mean that less attention should be given to human health and safety precautions; men will still have to enter the working environment to service and maintain equipment, and to reteach the robot when a new component is to be introduced. Indeed, when a robot is used, additional safety precautions will be needed. For instance, the robot must not be capable of generating a spark either from the mechanism itself or from the mechanism interacting with the environment. Precautions will need to be taken to prevent the build-up of static electricity by providing adequate earthing arrangements. Sensing devices should operate with low voltages and currents to minimise the risk in the event of a breakdown.

In some cases it may even be necessary to use robots that are motivated by hydraulic power rather than electric power. The control console will nearly always have to be situated well away from the robot itself, again to minimise the risk of spark or some form of electrical ignition of flammable vapours. The precautions necessary against erratic mechanical movements need not be so stringent. At worst, something other than the component will be sprayed, which is wasteful and inconvenient rather than unsafe.

Another application which illustrates different problems associated with work processes is that of forging. In essence it is a simple task to perform. A hot billet of metal is taken from a furnace and placed in a die. The die is located in a press and when the press is operated the billet is pounded into the correct shape for subsequent machining. Once the forging has been completed the billet is removed to a storage container for transportation to the next workstation or to the stores. When a man controls this process he provides the effort to move the billet from the furnace to the press and subsequently from press to hopper. He will also trigger the press when the billet is correctly in place.

In performing this operation a man is exposed to sparks and radiant heat from furnace and billet. The dangers of presses and often white-hot metal are self-evident, as are the safety precautions (e.g. safety interlocks and guards on presses, thick leather aprons, gloves and hats for the operators, shaded goggles or visors). When a man is replaced by a robot the problems from sparks and radiant heat are lessened as men can be removed from the immediate environment of the furnace.

Offsetting this immediate advantage is the difficulty robots have in handling hot billets that might be in any position within the forging jig. Unless the billet is forced into a fixed position with known orientation a robot cannot easily pick it up and move it to other parts of the forging process, or to other manufacturing processes. To accomplish this orientation a secondary process may have to be included within the manufacturing cycle.

These intermediate processes certainly reduce the robot 'fumble', but they do not lessen the danger should a fumble occur. Ironically it is one of the major reasons for using a robot for this transfer operation (i.e. its speed of movement) that causes a problem. When a man drops a billet of hot metal it is only likely to fall on the floor, but when a robot drops a billet it can make a lethal projectile. Clearly in this latter case, adequate safety screens to contain any such projectile would be very important. Other precautions such as safety gates on the press may still be needed to ensure the robot manipulator is clear of the press before it operates.

These examples illustrate two fundamental points:

(a) the safety systems of the robot and the work process have to be integrated, and
(b) using a robot system usually requires more safety precautions, not less.

With these points firmly established it is now worthwhile to consider the nature of possible robot malfunctions, given that they affect the magnitude, as well as the nature, of the safety problem in a working situation.

## 4.4 Robot system malfunctions

When robot systems malfunction, the causes of the malfunction can be many and varied, but by and large they manifest themselves in relatively few ways. The singlemost significant feature of any malfunction in a robot system is erratic behaviour, or more simply unexpected movements. These may be grouped as minor deviations from the taught path of operation, and major deviations from the taught path of operation.

To a trained technician the significance of these symptoms is great. Minor deviations or erratic movements are indicative of a system that is going out of alignment, rather like a car that is in need of tuning. The remedy for this kind of problem is simply to realign the control system and to check the mechanical integrity of the system (e.g. tighten loose bearings, check for worn components or couplings, etc.). If neglected it is possible for these minor problems to take on a more serious nature and so, regular maintenance is absolutely essential.

| *Symptoms* | *Possible causes* |
|---|---|
| Large involuntary twitch (sporadic) | Software corruptions<br>Loose mechanical coupling<br>Worn motor parts<br>Motor malfunction<br>Loose positional sensor<br>Electronic component/servo motors breaking down<br>Loose or broken electrical connection |
| Large deviation from taught path | Breakdown in control circuit<br>Software corruption during teaching<br>Loose bearing or coupling |
| 'Freezing' during work cycle | Power breakdown<br>Control circuit breakdown<br>Process malfunction<br>Computer breakdown |
| Robot moves at maximum speed to the limit of its movement and 'freezes' | Mechanical shearing<br>Motor breakdown<br>Control system breakdown<br>Computer breakdown |

*Fig. 4.2 Chart illustrating the relationship between some symptoms and their possible causes*

Major deviations or large erratic movements are usually symptoms of something far more fundamental going wrong. They can be generated from any part of the system and generally signify a major disruption or dislocation. Figure 4.2 illustrates the relationship between some symptoms and their possible causes. Charts such as these would be used by technicians and engineers for troubleshooting exercises and naturally the symptoms and causes will change from robot to robot.

Perhaps before proceeding it is worthwhile stressing that these are all very unusual events. To put them in perspective it must be realised that robot systems have a working life of between five and fifteen years without correct maintenance. A major disruption of the type described in the chart might only occur once or twice in the whole of the robot system's life.

The problem for the designer of the robot application system is that such rare events have to be anticipated. In the majority of cases the solutions to these problems are provided within the robot system itself. As the power of the computers used by most manufacturers of robot systems is high, it is possible to devote some of it to safety measures. In some cases computers use their programmed intelligence to scan the robot system to ensure that everything is as it ought to be. Should a deviation be noted (e.g. a large unprogrammed motion) then the computer can be programmed to take the appropriate remedial action, the nature of which must be matched to the application. In the case of paint spraying the remedial action might be a careful withdrawal to a 'home', safe position. With paint spraying the danger caused by unexpected motions is likely to be less than in the forging situation where the remedial action might be to freeze the robot and close down the whole process.

## 4.5 The human element

With automated systems people tend to be less exposed to the danger of work processes than in manual systems. But nonetheless a human interface is nearly always required even in ostensibly fully automated systems. With robot systems people become directly involved in at least three distinct situations: during programming, during repair and during maintenance. When these situations occur they are only dangerous to the extent that the person involved does not know how the robot system works, and what has to be done in the event of an emergency.

Simply because similar machinery has been tended in the in the past by an operator or maintenance technician it does not mean that they have an adequate level of knowledge to cope with a robot system. In short, people who operate, program, repair or maintain robot systems should not be allowed to perform these functions unless they have received adequate training. Since nearly all robot system manufacturers offer training free, or at a nominal price, there is every incentive to fulfil this basic obligation.

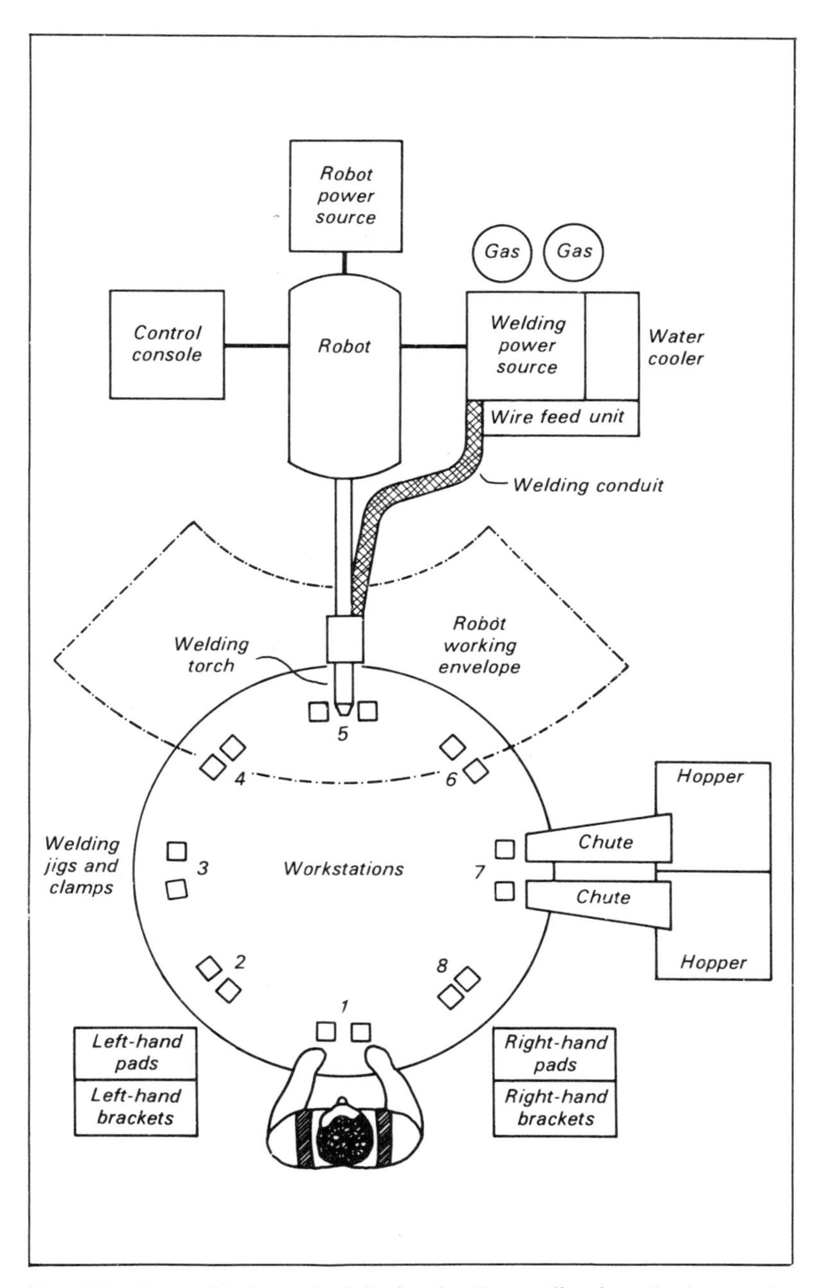


*Fig. 4.3 Example of a typical design for the application of robots to the process of arc welding (without any safety designed-in)*

## 4.6 Integration of safety aspects

Probably the best way of seeing how these separate aspects of safety are integrated in practice is to consider an example. A good example that embraces many of the problems of designing safe systems is that of applying robots to the process of arc welding.

Figure 4.3 shows a typical design for such an installation (without any safety designed-in). The operator loads the left-hand and right-hand pads and brackets on to the jigs in front of him. These jigs are open until they reach workstation 2 where they will be automatically closed. While the operator is loading the pads and brackets at station 1, the robot will be welding another set at station 5. When the loading at station 1 and welding at station 5 are complete the table will index through 45° and the previously described autoclamping will take place at workstation 2. As the indexing operation has been completed so a welded set of pads will be ejected at workstation 7.

With this arrangement the robot will not weld when the work table is indexing between workstations. (Note, workstation 1 is the loading station, workstation 2 the autoclamp station, workstation 5 the weld station, workstation 7 the auto eject station.)

To make the example realistic assume that the work to be completed is as in Fig. 4.4. There are two brackets, one left-hand and one right-hand. Six welds are required, three on each pad. There are two fillet welds 20 mm long and a plug weld located in the position shown. It may be assumed that the pads are to be welded in pairs and that the total cycle time has to be less than 30 s. The material is 16-gauge mild steel, previously prepared to an overall accuracy of ± 0.5 mm.

To avoid unnecessary technical discussion assume that the robot system has been bought as a turnkey package and includes welding power source, wire feed unit, welding conduit, water cooling unit and water-cooled welding torch. The table, and clamping and eject mechanisms will also be bought as a customised turnkey package.

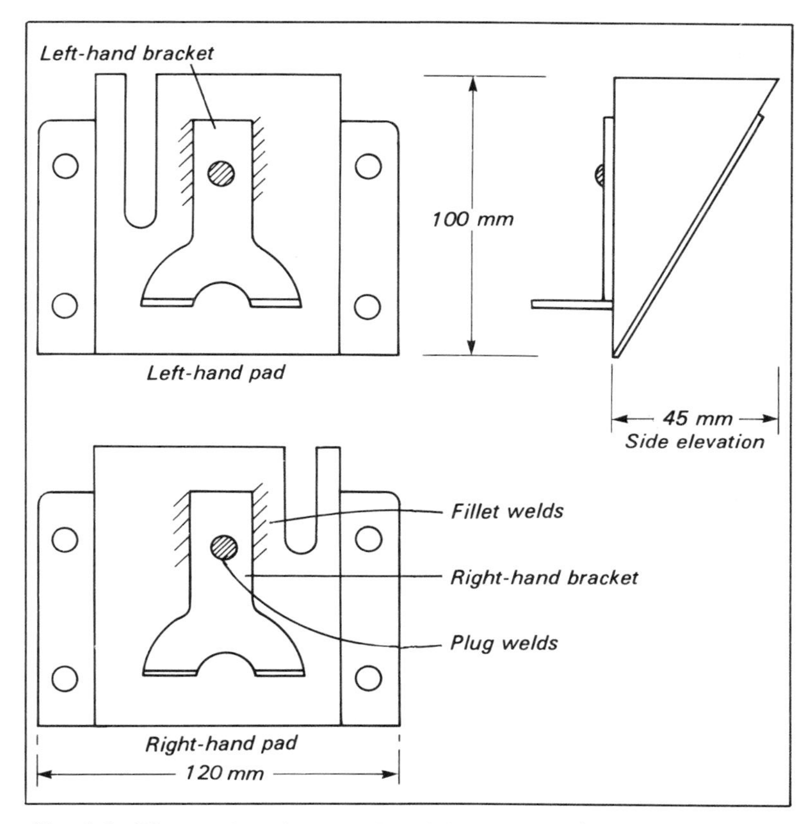


*Fig. 4.4 The work to be completed (i.e. the welding of the left- and right-hand pads) in the operations described in Fig. 4.3.*

Considering only the safety of such a system a number of general issues are immediately raised:

- The operators hands need to be kept away from the autoclamps when they operate to prevent fingers, etc., being trapped.
- The operator and people in the vicinity of the operation need eye protection against arc flash.
- The operator must not be able to be dragged around by the table when it indexes to the next position.
- People must be prevented from walking into the general work area during operation.

- If the robot chosen were of the hydraulic type the operator would have to be protected against the possibility of hydraulic rupture.
- Pads and brackets must be in position to prevent the jigs from being ruined.
- Ventilation will be required to prevent the build up of noxious fumes.
- Manual emergency stops will be required to close down the system in an emergency.
- Adequate precautions will have to be taken in the event of a fire (particularly with hydraulically operated robots).

The safety features that will be required within the robot welding system are:

- The welding system has to have safety interconnections that cause the arc to be extinguished in the event of:
  —water cooler failure
  —power unit (welding) malfunction
  —wire feed malfunction
  —a disruption or failure in gas supply.
- The welding system would have to be connected to the robot to ensure that if any of the above occur then appropriate alarms are sounded, and the robot withdraws to a safe home position until the error or breakdown is corrected.
- Any manual 'emergency' trigger will switch off the welding equipment and put the robot into a safe state.
- Any large positional errors of the robot should be detected and appropriate suspension of welding and table movements implemented.
- If the robot is powered by hydraulics the hydraulic hoses should be shielded against sparks and abrasion. Should a rupture occur then the hydraulic fuel should have flame retardant properties (e.g. quintolubric-based fluid).

If the robot is powered by electric motors then the cables that supply power will have to be protected against sparks and abrasion. This protection will have to conform to the relevant safety regulations.

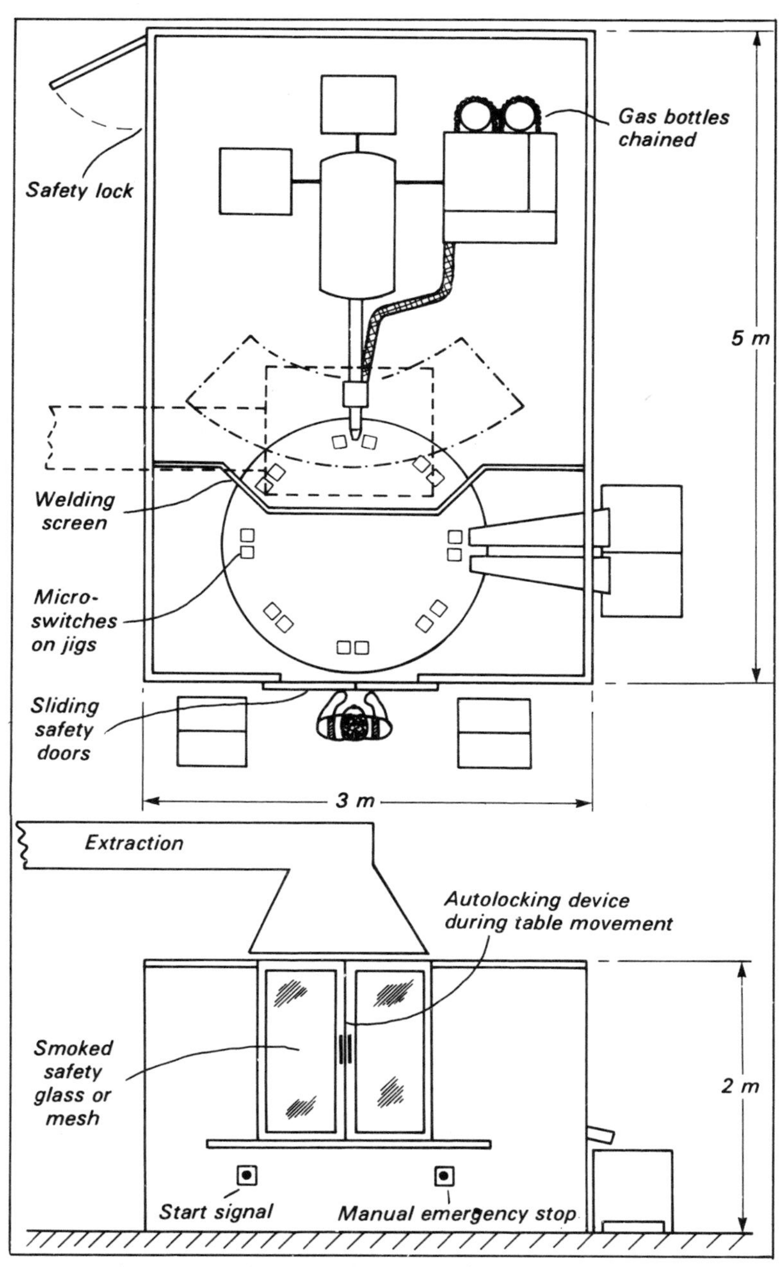


*Fig. 4.5 Integration of safety features into the design example for the application of robots to the process of arc welding*

Clearly this list is not complete, but these are all substantial requirements. Many of these features are often included as a matter of course with customised systems of the type previously described. If they are not included and are required, then they will have to be specified when the turnkey system is ordered.

If these features have been implemented then the next range of problems to be addressed are those previously listed as general problems. An examination of the general layout shows that simply by enclosing the whole operation in a solid barrier 2 m high would solve many problems. Naturally the operator would need to work outside the enclosures.

If the sequence of operations in Fig. 4.3 is now examined it can be seen that the robot will be welding when the operator is loading, and the table will be ready to index once the welding cycle has been completed. Both of these present safety problems: in the former case the operator can be protected by placing a welding screen between the welding station and the operator; in the latter case it is necessary to inhibit movement of the table and operation of the clamps until the operator has moved clear of the loading station. A gate system will ensure that this is accomplished.

Locking the gate electromechanically until indexing has been completed will ensure that the operator will not try to short-circuit safety precautions. The jig can be protected against accidental welding when the pads are not in position by having microswitches that operate when the pads are properly located. This of course would not need to be integrated with the logic that drives the index table.

Taking these precautions will ensure that the work process is reasonably safe; but what of the problems associated with programming, repairing and maintaining the system?

Once inside a safety enclosure the operator or maintenance technician is at risk. The major problem is that the person has to share the enclosure with the robot. This situation is inevitable and therefore it is important to ensure emergency stop buttons are placed within easy reach of the operator or technician. These may be fixed or mobile, or a combination of both. The fixed emergency stop buttons would need to be on each piece of equipment and around the perimeter of the enclosure; the mobile buttons attached to, say, the programming pendant.

Going back one stage, it is necessary that entry to the robot enclosure is via a safety interlocked door. The keys to the interlocking devices should be kept safely with someone in authority. In this way only trained and authorised personnel can gain entry.

Once inside the compound an operator or technician would require a full range of safety apparel for the normal welding environment, such as a welding mask, flameproof overalls, safety shoes or boots, chrome leather gloves and possibly a chrome leather skull cap. Also, the welding and indexing table equipment needs correct earthing, gas bottles prevented from toppling, and arc gasses need either local or general extraction. (Figure 4.5 illustrates how these safety features can be integrated into the already proposed system.)

Although the examples chosen in this chapter have been of single robot applications, the principles apply equally to multiple robot applications. The practice of making working robots safe is not complicated, and as was stated earlier the majority of measures are common sense and related to safety practices already being used.

# *Chapter Five*
# People and new technology

THE PRESSURE to introduce robots or automation into a factory usually comes from the market place. Goods have to be sold at an economic price and therefore if profit margins are to be adequate, manufacturing efficiency must be high. In some areas of the world labour is so cheap that there is little incentive to automate. But in the developed countries labour is expensive and the solution to manufacturing efficiency is found in the utilisation of capital equipment.

The circumstances in which manufacturers have to introduce robots and automation will vary greatly. Some companies will be unionised, and of course management styles will vary from autocratic to democratic. Whichever conditions prevail the task of management is to ensure that robots, when installed, are put to work as quickly as possible.

While management incentives to adopt robot technology are usually quite clear the same is not always true of the workforce who do not necessarily share management's aspirations. For many workers robots represent the unknown which will be regarded with suspicion. Thus there is always the potential for conflict. Whether or not the potential conflict becomes a reality depends entirely upon how well management perform its task.

It is a fact that many workpeople are suspicious of robots and automation. Robots pose a threat and represent a large element of the unknown. Whether or not it is management's intention to use robots to replace people, the workforce tend to assume the worst and become very defensive. Two basic courses of action tend to follow: (a) the workpeople accept and adopt the robots, or (b) the workpeople ignore and reject the robots. It is difficult to predict which of these will be the outcome. However, it is possible to influence the workpeople if managers suspect that a negative reaction may result from their using robots. Indeed it is arguable that whether or not management are worried about this problem they would be well advised to 'head-off' trouble. In practice this is relatively simple to do and managers who have introduced robots (and this includes those in the case studies in Part III) have all adopted a fairly simple approach:

(a) to inform their workpeople at an early stage about what has been planned and when things are going to happen,
(b) to educate their workpeople as to what robots will mean to the company and to themselves,
(c) to train their workpeople to use the robot effectively and efficiently, and
(d) where relevant, to negotiate with the relevant trade unions.

## 5.1 Informing and educating the workforce

The importance of workpeople having the right attitudes towards robots and robot systems cannot be over emphasised. This point is simply illustrated by the author's own experience of a robot system that produced good work on one shift and bad work on the following shift. The robots were the same, the materials were the same, the process was the same, only the people were different. On the first shift the workers had a positive attitude towards the robot system and looked on it as making their job easier. On the second shift it almost seemed as though the workers did not want the system to work.

These differences in attitude were reflected in the percentage of substandard components produced on the two shifts. On the first shift the reject level averaged 3% and on the second shift it was 17%. The root of the trouble was eventually traced back to the early stages of the introduction of the robot system. The management took the trouble to explain the new system in detail to the first shift and assumed the second shift would get the information by word-of-mouth. To some extent this did happen, but the information that had been passed on had been subject to a 'personal' interpretation by a particularly antagonistic shop steward.

Managers need to take as much trouble over launching a new robot system in the factory as they would for launching a new product on the market. They must present the facts clearly, and probably the best starting point will be to imagine what questions the workpeople will be asking:

Why does the company want to use robots?
How will they be used?
What changes will occur?
How will jobs be changed and who will be affected?
Who will be trained and how will they be trained?
Will anyone be made redundant?
If people are not required will they be redeployed?
What will happen to the existing work system?
Are the unions being consulted?

There will of course be other questions and it seems to be the best policy to answer questions as frankly as possible. Above all, be prepared.

The type of presentation required will vary in detail depending upon who is being informed. People in departments directly involved with the robot system will require a much higher level of detail than others. The management's presentation can be backed-up by support from suppliers of the robot system. They often have slides and films that are easily understood by the workpeople, and as they are used to this kind of 'selling' activity usually do a good job, especially when properly briefed by local management on sensitive issues.

In short, management must use any and every technique available to ensure that their workforce are 'sold', and kept sold, on what the company is trying to do.

## 5.2 Training the workforce

With new technology there is an inevitable need for new skills. Who needs these skills and how are they to be acquired?

The range of skills required will vary greatly depending on who is involved. By and large the greatest demand for new skills will be among the people directly involved with the new system: operators, foremen and supervisors. At the operator level the skills required to use the robot system are evident. They will need to know how to stop and start the robot system in both normal and emergency situations. If the robot system has to be programmed by the operator (as is the case in a process such as fettling) then he must be given full training in what to do and what to expect. Indeed this is not a daunting prospect as the majority of robot systems are designed for ease of use. Beyond these operational skills is safety training, and this is particularly important when operators come close to the robot.

Operational skills will be greatly enhanced through practice, and any training course must have plenty of 'hands-on' experience. Usually initial training is done off-site and it is as well to allow adequate time after on-site installations for more practice before serious production begins. The majority of robot users find that it takes about two months before operators are fully competent, by which time the robot system will already be in use. It is wise to ensure that as many operators as possible are trained to use the robot system, as this reduces the company's vulnerability to sickness and labour turnover, and prevents any one person becoming a prima donna.

The level of management immediately above the operators must also be thoroughly familiar with the robot system and know as much as the operators. This is essential if their authority is not to be undermined and their control of workpeople and work process supported. Beyond the basic skill level needed to operate the system the foremen or supervisors must be aware of the performance parameters of the system. This will enable them to monitor the quality and output of the system and where necessary feed information back into the manufacturing system about product or process design. As with the operators the training of foremen or supervisors should be a combination of on-the-job and off-the-job experience.

The training required by middle and senior managers is of a more conceptual nature. Since the role of these managers is to plan and implement change in the company it is clear that their training should begin before the decision to use robots is made. This training should continue throughout the introduction of the new robot systems. The following plan is a useful guideline:

*Predecision phase*—attendance at seminars, exhibitions and lectures to become familiar with what robots are and what they can and cannot do. Often at these occasions it is possible to talk to other managers of similar standing who have already implemented robot systems. Some larger companies employ robot consultants at this early stage to conduct in-house training. This training should include familiarisation with robots and give individual managers opportunities to discuss the implication of robots at a detailed level as related to their own functions.

*Decision phase*—the basic training given previously must be reinforced during this phase. Many companies find that a team approach to making the decision, with expert help, provides a good basis for highlighting any individual or organisational problems.

*Post-decision phase*—during the post-decision phase it is important to keep middle and senior managers up-to-date with progress. Specific training from the vendors of the robot equipment and systems designers is particularly useful at this stage.

Last and by no means least is the training of the company's technicians. The training of these people must be conditional upon the level of skill and education of the existing maintenance people. Those who have a background in maintaining electrical and electronic systems will probably find little difficulty in coping with the maintenance of robot systems. However, the same will not necessarily be true of the maintenance technician who has only experience in maintaining mechanical systems. In the latter case it would probably be as well to seriously consider employing someone with a suitable background for training.

Assuming that the right people are available nearly all the training for maintenance will be done off-site. This training will include fault-finding diagnoses for each part of the robot system. The technician should be provided with all the documentation and circuits to keep the robot system going. It is also usually necessary to buy special tools and spares to ensure that the training can be put to use in the factory.

Fortunately the majority of robot systems can be repaired in a straightforward way with replacement electrical or mechanical components and complete printed circuit boards. Complex system problems usually require the attention of an expert from the robot system supplier. Some companies who are not concerned with a fast response time often subcontract all maintenance to the system suppliers.

## 5.3 Trades union negotiations

The relationship between management and trade unions, often called industrial relations, is a notoriously tricky area to write about. This is largely because of the variety of situations that prevail within and between companies and countries. Each situation is different and influenced by the people concerned and by the history and traditions of local and national organisations.

By and large the majority of managers find a workable arrangement with their workpeople based on a mutual recognition of what is practicable and what is not. For some managers this is achieved through consultations, for others it is achieved through confrontation. Whatever the situation the introduction of robots makes a change to the status quo and is therefore a possible subject of negotiations.

Before taking a closer view of these negotiations it is interesting to examine the general situation vis-à-vis employers' associations and trade unions. As far as employers' associations are concerned there is little to be found in the way of a common policy. This is probably a reflection of the value that employers place on autonomy and an unwillingness to escalate employment policy beyond the workplace. To some extent this is understandable because this is the level at which they are the strongest. On the other hand trade unions are traditionally weakest at factory level and their strength lies in their ability to escalate negotiations to a level where their greatest strength lies—the local and national officers.

Because of these differences in traditions it comes as no surprise that many of the trade unions throughout the world have evolved quite comprehensive policies to the many issues that are facing them. In particular they have given a great deal of thought to the introduction of technology and automation/computerisation in general.

These policies are usually translated into guidelines which are given to local trades union negotiators. In the UK it is the TUC who have given much of the lead in the formulation of policy guidelines in relation to new technologies. Readers particularly interested in the TUC Guidelines to New Technology Agreements should contact the TUC Research Unit in London. Otherwise, specific trades union guidelines may be obtained from the central offices of the trades union concerned. These policy guidelines have served as a focus for many of the member unions in the TUC and have variously been translated into "New Technology Agreements". These translations reflect the nature of the union membership and range from office automation to factory automation.

Clearly it is both in managers' and workers' interests to ensure that the introduction of robot technology is a positive process. If agreement can be reached quickly and amicably then all parties will be satisfied. In the majority of cases this is done through the application of common sense and within negotiating procedures. Occasionally, because of local bargaining conditions, trades union representatives may actually table a "Model New Technology Agreement." This will have been provided by the district officer of the union.

What do these New Technology Agreements contain? Appendix B is a substantial extract from just such a document. A superficial reading of this document could easily lead to the opinion that it is aggressively laced with demands and assertions. However, if a little more care is taken it can be seen that beneath this there is a spirited defence of the status quo while at the same time tacitly accepting the inevitability of change. Certainly there are some clearly stated long-term trades union objectives (a 35 hour working week, 6 weeks annual holiday, and so on) but in the melée of local bargaining these are often sacrificed in favour of more substantive issues such as additional skill payments, retraining and re-deployment.

Faced with a document such as the model agreement it is managements' responsibility to evolve a clear strategy for the local factory situation. To do this it is important to decide from the onset what elements are going to be essential to the introduction of the robot system and to making it work in a viable way. In the last analysis both managers and workers should be expending their energy on making the new technology work, not in continual negotiations.

Throughout the world the spread of robot technology has been, and is, as inevitable as the spread of computers. To ignore it is to ignore progress and in manufacturing terms this is often a shortcut to economic catastrophe. By taking common sense attitudes with their workforce, managers have successfully introduced robots in many different working environments—usually making both managers' and workers' jobs a lot easier and also increasing productivity.

In the final analysis managers control the working environment and they get the industrial relations that they deserve. Poor planning, bad selling, bad negotiations and inadequate training can all be the cause of poor robot systems performance. Whenever this has been the case managers blame the robots or their workers, rarely themselves!

# *Chapter Six*
# Finance and robot systems

TO HAVE reached a point at which a company has begun to consider the introduction of robot systems usually means that some change is in order. Going through the processes of company analysis advocated in previous chapters will often show alternative solutions to existing problems that can be implemented without using robots at all. If these solutions are cheaper and more cost effective than using a robot system, all well and good—for in the last analysis the best solution should always be sought. However, if these alternatives are not acceptable, for whatever reason, then it is important to obtain a realistic appraisal of the financial implications of buying a robot system.

The only justifiable case for purchasing a robot system is when it will achieve, or help to achieve, the company's overall objectives. It does not matter whether these objectives are expressed as a target for growth or of increased efficiency, the only concrete measure at the end of the day is the contribution the robot makes to the bottom line of the company's balance sheet. Managers and engineers do tend to get carried away with the beauty of 'their' system and forget this fact. While in this frame of mind it is not unusual for them to over estimate the benefits of a robot system and under estimate the costs.

With this last point in mind this chapter begins by identifying the cost structure associated with the purchase of robot systems. There follows a short discussion of tax and government grants as ways of defraying some of these costs. Finally a well-established method of investment appraisal is outlined.

## 6.1 Main cost areas

The costs associated with any robot project can be categorised as follows:

Manufacturing equipment.
Systems engineering.
Structural modifications and ancillary equipment.
Safety.
Personnel training and recruitment.
Computer hardware and software.
Project management.
Miscellaneous.

Clearly it is difficult to be precise about exact costs simply because of the variety of robot applications. However, it is possible to be fairly precise about the factors which affect the magnitude of these costs; and where practical an indication of the range of cost is given as a percentage of the basic manufacturing equipment.

**Manufacturing equipment costs.** Of all the areas of costing this should be fairly straightforward simply because the work completed in Chapter Three will take the reader to the point at which each piece of equipment can be individually costed. Even if the system is to be built on a subcontract basis the exercise of individually pricing each piece of equipment is a useful check against under or over costing. Any development costs associated with new equipment may add some uncertainty about the precision of the costing exercise. When this is the case a scaling factor may need to be introduced. The size of the scaling factor should reflect the level of uncertainty about the outcome of any development activities. Generally the cost of manufacturing equipment can be categorised:

*The robot system*—will comprise of the manipulator with its associated control system and motive source. Sometimes the choice of robot may be influenced by factors other than price. Reliability, compatability, availability, package deals and other factors may all play an important role.

*Package enchancements*—although bought as a standard package the majority of robot systems will require some additional enhancements. These could be in the form of additional memory capacity, extra interface channels, special control features, and so on. All of these extras add to the basic cost of the system. Usually such additional items are expensive and if forgotten can be a source of embarrassment when unbudgetted bills arrive.

*Grippers or work handlers*—sometimes these are supplied with the robot as a standard package and may not be suitable for the work to be done. More often than not grippers are not included and will have to be individually designed for a particular job. Grippers are rarely simple mechanisms and often require some additional power source. This kind of one-off engineering can be expensive and the price is largely dependent upon size and complexity. For a simple small gripper the cost may only be 1% of the robot system costs; for a complex gripper with sophisticated sensing the cost could be as high as 20%.

*Table or work positioners*—this equipment is very important in robot applications such as arc welding. The equipment will be required to move workpieces into positions where the robot can do its work efficiently and effectively. In general there is a relatively simple relationship between the size and weight of the component to be moved and the cost of the table or work positioner. The larger and heavier the workpiece the more expensive the equipment will be. In some cases where special work manipulation is required (e.g. in the nuclear industry) it is not unusual to find that the manipulator may even be more expensive than the robot system. When standard tables or positioners can be used their price is usually in the region of 10–15% of the robot system cost.

*Jigs, fixtures and tools*—nearly all robot system applications require one of these items. Normally they will have to be specially designed to work with robots. Taken as a whole the expected costs for these items is between 15 and 20% of the robot system cost.

*System interlocks and interfaces*—the work required to integrate any system will require specific interlocking and interfacing electronics and devices. This work can involve the designing, building and testing of quite complex systems and may even involve the use of microprocessors. Whether simple or complex, the components that are used in these design activities are readily available. Because of this, development time is relatively quick. In large multirobot systems the cost of interlocking and interfacing can be as high as 12–20% of the cost of the robot system, for more straightforward systems the cost is typically 10–15%.

*Process equipment*—the many examples of robot applications in this book illustrate the variety of process equipment that could be required. The range is enormous: conveyors, ovens, pallets, hoppers, feeders, magazines, millers, lathes, grinders, and furnaces, to name but a few. Because of this variety it is impossible to accurately establish a range of cost. All that can be said is that it will be totally dependent on the application.

Taken as a whole, with the exception of process equipment, it is often found when these costs are totalled that they are about twice the cost of the basic robot system.

**System engineering costs.** The major factor that dictates the cost of system engineering is labour. As an activity it involves engineers and technicians whose time is very expensive. However, having made this point, it is always worthwhile paying for the best as this is usually a form of insurance. Many robot systems have taken longer to get into production than they ought to because of bad system engineering.

Whether using internal or external resources, always ask for a detailed breakdown of the times that are allocated for the following major tasks:

Buying the equipment for the system.
Testing the equipment against manufacturers' and system specifications.
Progressively assembling the system and testing/commissioning interlocks and interfaces.
Testing the whole system against system criteria.
Breaking down the system for transportation.
Transportation and insurance.
Reassembly.
Retesting.
Running-up the system into full production.
Travel and expenses.

(Note, if the system is assembled and tested on the buyer's site then the last five items need not be detailed.)

As a general guide the expected system engineering cost is 10–15% of the basic system cost.

**Structural modifications and ancillary equipment costs.** This is all work that has to be carried out before the robot can be installed. Sometimes when space is freely available—perhaps in the case of a new factory—the work associated with this activity may simply be the provision of services such as electricity or ventilation. In other circumstances it can involve the removal of old equipment, resurfacing floors, changing service routes, adding ventilation or extraction equipment, knocking down old walls or partitions and building new ones—in short, a lot of work and a lot of expense. In the first case the cost of this activity will obviously be very low; in the second case the cost could literally run into thousands of pounds.

**Safety costs.** Arguably safety costs should be included in the equipment section. However, in view of the importance of this subject it is worthy of separate attention. The exact nature of the safety requirements will have to be designed for each and every application. Once designed they will require the approval of the appropriate safety inspector. The main elements of the safety system will be:

*Safety enclosures*—which can be wire screens, solid screens, or a combination of both. These will be individually designed for every application but can often be assembled from standard units.

*Alarm devices*—typically emergency stop devices or claxons that warn of equipment malfunction or operator distress.

*Operator protection devices*—used to ensure that the operator is well clear of the equipment when it is working. Pressure mats, light curtains, guards and rails are all examples of specific devices used to protect operators.

*Safety training*—this must be given to everyone who works or comes into contact with the robot system. It often involves time away from the worksite.

All of these elements collectively form the safety system. Typically these will add about 10% to the cost of the basic system.

**Personnel costs.** The costs associated with the personnel aspects of implementing the robot system are usually found in the need for skill training, wage settlements, redeployment or redundancy, and recruitment.

*Skill training*—this is usually provided free by the robot supplier, but some allocation should be made for the operator's expenses whilst attending the training course and also possibly for the loss of work in the operator's absence. Bearing in mind that in the manufacturing industry operator turnover is normally in the region of 10–15%, provision must be made for the training of new operators who are recruited after the system purchase.

*Wage settlements*—these may be specifically related to the people who are operating the robot system, or alternatively the new robot technology may be used to negotiate a general 'new technology' pay rise. As wage bargaining is essentially a local activity it is impossible to predict with any certainty what level of additional cost, if any, will be involved. However, it is heartening to find that in many companies trade unions do not automatically demand pay rises and increasingly are actually encouraging management to use new technologies in order to project the long-term future of jobs.

*Redeployment, redundancies and recruitment*—all of these items are cost associated. Redeployment usually means some form of retraining and the resultant temporary loss in production. The training may have to be done away from the factory if the company does not have any training facilities of its own, and this can be quite expensive. Sometimes there are grants available to offset the costs of this training and the relevant government department should be contacted. Redundancies, if caused by the introduction of robot technology, is a far more complicated issue to deal with. While it is true that there are legal minimum payments to be given that are related to length of service, some additional payments may be required to buy an element of acceptance from the remaining labour force. Recruitment is usually required in companies that do not have an existing technical infrastructure. This represents an addition to indirect costs.

By and large the costs that are associated with personnel tend to be rather unpredictable. In the melée of local bargaining many issues become shrouded in political overtones. Issues that apparently have been noncontentious suddenly take on different dimensions. All that can be done in practice is to face issues as and when they arise.

**Computer hardware and software.** For many robot applications there will never be the need for any computers other than those that are supplied to control the robot. However, the general trend in manufacturing systems is to use many robots in manufacturing cells. In such instances these robots are often controlled by a central computer.

To price the computer hardware and software they are best considered separately. A typical computer hardware configuration for controlling a manufacturing cell could comprise of a central processor unit, a core store (working memory area), bulk storage media (disks, magnetic tape, etc.), interface modules, visual display units, graphics terminals, and a printer.

The size and speed of each of these elements can only be determined by a detailed feasibility study (see Chapter One). It is not unusual to find that the cost of computing hardware is about 50–70% of the basic robot system cost.

Computer software costs are related to the complexity of the system. By and large software can be divided into three categories:

(a) Software packages that contain the standard computer language that the system is to be written in, and the software that contains the operating system of the robot.
(b) Standard software packages that have been developed by software houses for specific functions (e.g. production control systems).
(c) Customised software that is specifically developed for one system problem (e.g. the software for a flexible manufacturing system).

Items (a) and (b) are usually off-the-shelf units that have a specific cost and are reasonably priced simply because their initial costs have been amortised over many sales. Item (c) is inevitably the most expensive of all software costs. In sophisticated applications this special software can be two or three times the cost of the basic robot system and it is always necessary to have expert assistance in the assessment of any proposal. With all computer software it is best kept simple and only have what is required. However, with both hardware and software it is always best to make allowances for future expansion.

**Project management costs.** There are very few projects that have been successfully implemented without a project manager. Such a person's role is to:

(a) plan the activities involved in the project from inception to completion,
(b) ensure resources are available to complete the tasks involved in the project (this may also extend to extensive monitoring of subcontract resources),
(c) communicate with the people involved in the project to make sure the project targets are clearly understood and that all are motivated to achieve these targets, and
(d) control the project to ensure that all work is completed to time, specification and cost.

Associated with these activities are the inevitable costs for wages, secretarial help, expenses, and so on. For budgetting purposes it is as well to allow about 10–15% of the basic robot system costs.

**Miscellaneous costs.** The major costs which do not easily fall into the previous categories are survey or feasibility costs, maintenance and spares costs, and learning costs.

Throughout this chapter there has been reference to the need to consult experts if the company has none of its own. Before starting any robot project it is advisable to have experienced people to do the feasibility study. The cost and quality of this activity will be related to the time spent on the initial analysis and the quality of the people who perform the analysis. It is very much a case of 'you get what you pay for.' Management must decide how prescriptive they want the feasibility study to be. For a simple fact-gathering exercise about the existing system to be effective there have to be resources within the company available and capable of translating this information into a usable state.

Maintenance costs are inevitable once guarantee periods expire. Maintenance contracts for both hardware and software are usually about 10–12% of the gross capital cost for each respective item. Some companies will reduce these costs by doing their own first line maintenance. When this course is chosen then spares will have to be bought and maintenance technicians trained. In practice this latter course of action is best chosen when down time is very critical and a fast response is the most cost-effective way of managing the situation.

Finally there are the costs of learning about new technology. This has nothing to do with the cost of on-the-job or off-the-job training, but is concerned with the cost of mistakes, lost production and scrap that is inevitably produced when the company and individuals are learning about using robots. The size of this cost is difficult to predict with any degree of certainty but may be anticipated as a very small percentage of the total work going through the robot installation.

Overall the cost of the robot system bears a direct relationship with the complexity and size of the system. For simple systems the bulk of the costs are usually associated with the robot system as bought from the manufacturer, and with its associated hardware. With sophisticated systems the costs shift quite dramatically towards the nonrobot equipment.

## 6.2 Tax and grants

A full discussion of the subject of tax and how it relates at a detailed level to investing in robot systems is outside the scope of this book. Suffice to say that it is possible to offset some of the costs of purchasing the system against tax. As has been shown in the first section of this chapter there is quite a complex cost structure, and thus there is plenty of scope for 'creative' accountancy.

For tax purposes the expenditure on the robot system can be considered as falling into two categories: the revenue costs (associated with factors such as labour, materials and maintenance), and capital costs (largely associated with the purchase of equipment that appears as a company asset). Normally revenue costs are totally allowable for tax purposes as they are a business expense. They do not contribute to the capital structure of the company and so are allowable at the tax rate prevailing in any fiscal year. Capital purchases are treated differently, largely because the equipment, once purchased, has a nominal life over which it will depreciate in value. It is the amount of depreciation that attracts tax relief (see Section 6.3).

Along with tax relief the case outgoings required to purchase the robot system can be lessened due to the availability of government grants. By and large government grants fall into two categories: returnable and nonreturnable. Returnable grants are usually secured against the equipment purchased and repayment based on a levy on company profits over a period of time, or upon the sale of subsequent systems. Nonreturnable grants are usually for a lower percentage of the gross cost, but to offset this disadvantage they are not repayable. Generally nonreturnable grants are available through special government-devised schemes to encourage industry to adopt new technologies.

In the UK, grants are available through the Department of Industry and usually range from $33\frac{1}{3}$ to 50% of the total project costs. However, to qualify for the higher level of support the robot project usually has to have an unusually high level of innovation.

## 6.3 Investment appraisal

The sole purpose of any capital investment programme is to generate profits in the future. Normally the future is defined in terms of the life of the capital equipment purchased. For the most part, capital equipment is written down in the company's books over a period of five to ten years. The difficulty in looking to the future and trying to assess how good a particular investment will be is the changing value of money. To overcome these problems of inflation, accountants have devised a number of techniques, the most widely used being discounted cash flow (DCF).

DCF is based on the premise that it is possible to guess the rate at which inflation is running. In the UK in 1981 inflation was running at about 12–14% per annum, and many companies, erring on the side of caution, used an inflation figure of 15% for the purposes of DCF calculations. If the rate of inflation is expressed as $R$, then $(1+R)^{-1}$ gives the value of £1 in a years time, and $(1+R)^{-2}$ gives the value of £1 in two years time, and so on. Fortunately it is not necessary to perform these calculations as tables are available for various DCF factors. Using the example of 15% DCF the value of cash over a period of five years can be calculated using the following multipliers:

| Year 0 | Year 1 | Year 2 | Year 3 | Year 4 | Year 5 |
|---|---|---|---|---|---|
| 1.0 | 0.8695 | 0.7561 | 0.6575 | 0.5717 | 0.4971 |

If an investment involved an initial outlay of £1000 and generated an income of £300 per year for five years, the net present value of the proposition can be calculated:

| Year | 0 | 1 | 2 | 3 | 4 | 5 |
|---|---|---|---|---|---|---|
| Income | — | 300 | 300 | 300 | 300 | 300 |
| DCF | 1 | 0.8695 | 0.7561 | 0.6575 | 0.5717 | 0.4971 |
| Present value | 1000 | 260.85 | 226.83 | 197.25 | 171.51 | 149.13 |
| Balance | £-1000 | £-739.15 | £-512.32 | £-315.07 | £-143.56 | £5.57 |

If the assumed rate of discount is correct then this calculation shows that in year five a small profit is generated. So, if the company were looking for payback within four years then this investment should be rejected. Likewise it is unlikely that the company would expect a return on investment of 0.005%.

This is a simple example, but the principle is clearly demonstrated. In practice this technique must be applied to all capital cash flow calculations including those associated with tax.

To perform the calculations required for a DCF appraisal on a robot system the following must be known:

(a) the total capital cash outgoings associated with the robot system,
(b) the expected annual cash income generated by the robot system, or alternatively the amount of cash the robot system is expected to save per annum.
(c) the expected life of the robot system,

(d) the DCF factor to be used for calculations, and
(e) the rate at which the equipment is to be written-off.

Consider the following example:

| | |
|---|---|
| Capital cost of the robot system | £80 000 |
| Income generated by system | £35 000 per annum |
| Life of the robot system | 5 years |
| DCF rate to be used | 17% |
| Write-off rate | 30% per annum |
| Residual value of system | £5000 |
| Government grant (25%) | £20 000 |

*Stage 1*—The cash flow in relation to the prevailing rate of tax (say 48%) and assuming a maximum annual tax allowance of 30%

| | | | *Cash value at 48% tax* |
|---|---|---|---|
| *Cost of robot system* | | *80 000* | |
| *Less annual write-off (30%)* | *24 000* | | |
| *Less government grant (25%)* | *20 000* | | |
| | *44 000* | *44 000* | |
| | | *36 000* | |
| *Year 1 write-off allowance (30%)* | | *10 800* | *5184* |
| | | *25 200* | |
| *Year 2 write-off allowance (30%)* | | *7560* | *3629* |
| | | *17 640* | |
| *Year 3 write-off allowance (30%)* | | *5292* | *2540* |
| | | *12 348* | |
| *Year 4 write-off allowance (30%)* | | | *1778* |
| | | *8644* | |
| *Year 5 residual value* | | *5000* | |
| | *Balancing allowance* | *3644* | *1749* |
| | | | *£14 880* |

### *Stage 2*—Cash flow summary

| Year | 0 | 1 | 2 | 3 | 4 | 5 | 6 |
|---|---|---|---|---|---|---|---|
| Capital cost | −80 000 | | | | | | |
| Residual value | | | | | | | 5000 |
| Government grant | | 20 000 | | | | | |
| Write-off | | 24 000 | | | | | |
| Cash value | | | 5184 | 3629 | 2540 | 1778 | 1749 |
| Income | | 35 000 | 35 000 | 35 000 | 35 000 | 35 000 | |
| Corporate tax | | | -16 800 | -16 800 | -16 800 | -16 800 | -16 800 |
| Balance | £-80000 | £79000 | £23 384 | £21 829 | £20 740 | £19 978 | £-15 051 |

### *Stage 3*—Time adjustment of the cash flow figures

| Year | Net cash flow | DCF (17%) | Net present value |
|---|---|---|---|
| 0 | -80 000 | 1 | -80 000 |
| 1 | 79 000 | 0.855 | 67 545 |
| 2 | 23 384 | 0.731 | 17 093 |
| 3 | 21 829 | 0.624 | 13 621 |
| 4 | 20 740 | 0.534 | 11 075 |
| 5 | 19 978 | 0.456 | 9109 |
| 6 | -10 051 | 0.390 | -3919 |
| | | | £34 524 |

*Note, although the investment is written down over five years since tax effectively lags by one year the discount calculations are spread over six years*

From the calculations in Stage 3 it can be seen that the payback period of this investment is about 21 months and that the rate of return on the original capital invested is about 43%.

Throughout this chapter the emphasis has been on an accurate identification of costs. Once identified they can be offset partly through tax relief and sometimes through government grants. Then, by the use of DCF a realistic assessment of the investment in a robot system can be derived. In all cases it must be emphasised that the availability of grants should not be used to support an otherwise weak case. To draw all of this together Appendix C takes the reader through a complete financial appraisal of an assembly robot installation.

It must also be stressed that this chapter has taken a simple example. In practice many companies will choose to lease equipment rather than buy it, and other aspects of the financial management of the investment programme will be decided in line with current company financial practices.

# *Part III*
# Case Studies

# Introduction

THROUGHOUT the first two parts of this book attention has been paid to the practical aspects of making robots work effectively in the working environment. This, the third part of the book, is devoted to illustrations drawn from some of the industrial countries of Europe. This has been done quite deliberately to emphasise the international nature of robot technology. It also demonstrates that no matter what the industrial traditions of a country, the skills required to apply a robot are similar. Although these illustrations are quite brief, care has been taken to try to present a balanced view of the different companies.

All case studies follow the same basic format:

An introduction to the company.
The history of the company and the events that led the company to use robots.
The product or process on which the robot is used.
The robot installation and workcycle.
A brief description of the replaced manual system.
Designing, installing and running the system.
Labour relations.
Financial considerations.
Important learning points.
Future plans.

Naturally the level of detail in these sections does vary quite considerably. Seeming omissions are usually due to companies not wishing to have a particular issue aired publicly. This was especially true of financial information. In spite of these reservations the studies do give a view of robots 'earning their living' in real productive environments.

# *Case Study 1*
# Aspera SpA
## (Turin, Italy)

ASPERA is part of the components division of Fiat. The company's head office is in Turin and the robot system is located in the company's factory at Chieri to the south-east of Turin. Here, the company produces compressors for refrigeration and air conditioning units. The factory employs about 1000 people, a large percentage of which are associated directly with the production lines. This direct labour is classified as unskilled or semi-skilled and recruited from the surrounding agricultural communities. Both men and women work on the production lines, with the men taking some of the heavier work.

Aspera has a substantial proportion of the European market for refrigerator compressors.

### Company history

About 30 years ago Aspera bought the license to manufacture refrigerator and air compressors together with two- and four-stroke engines from Tecumseh Products Inc. of Illinois, USA. Aspera manufactured these products more or less unaltered for about 15 years. Then in 1969 Aspera decided to design and produce its own compressors. At the same time the decision was made to continue with the licensing arrangements for the petrol engines. Since then the company has extended its product line to include 0.5–10 hp refrigerator compressors and 0.5–15 hp air conditioning compressors.

In 1978 Aspera rationalised its product range and began to seriously consider how it might introduce new manufacturing techniques into the factory. This happily coincided with the evolution of a new robot by DEA of Turin and both companies undertook a collaborative venture to develop robot assembly

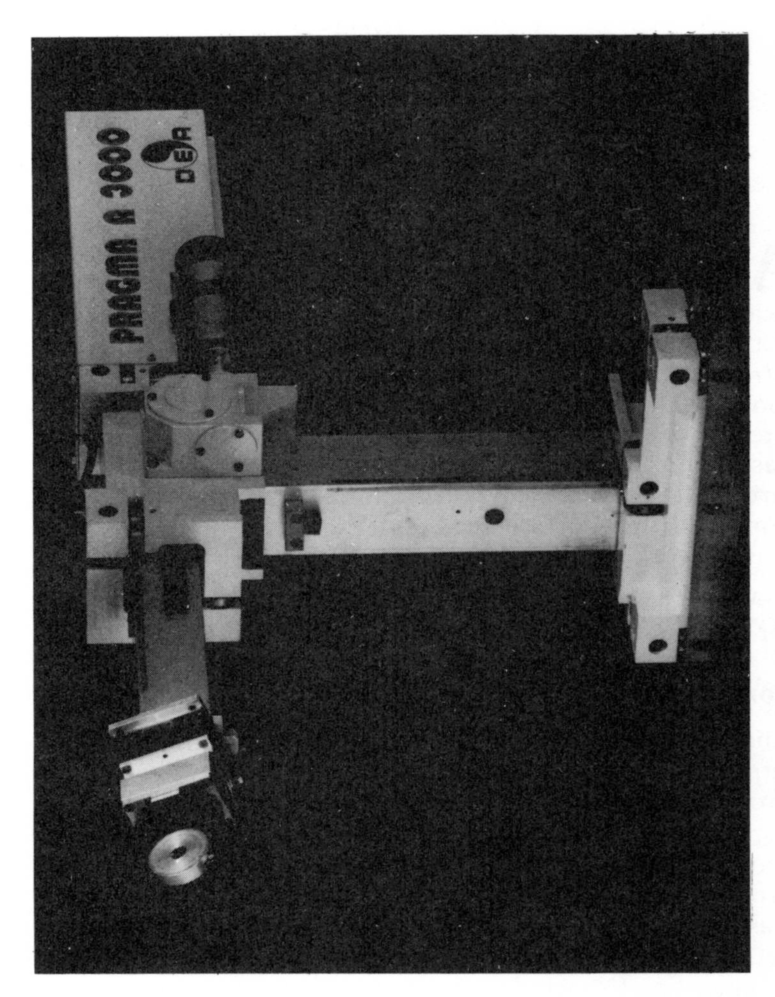


*Fig. 1 The DEA Pragma A 3000 industrial assembly robot*

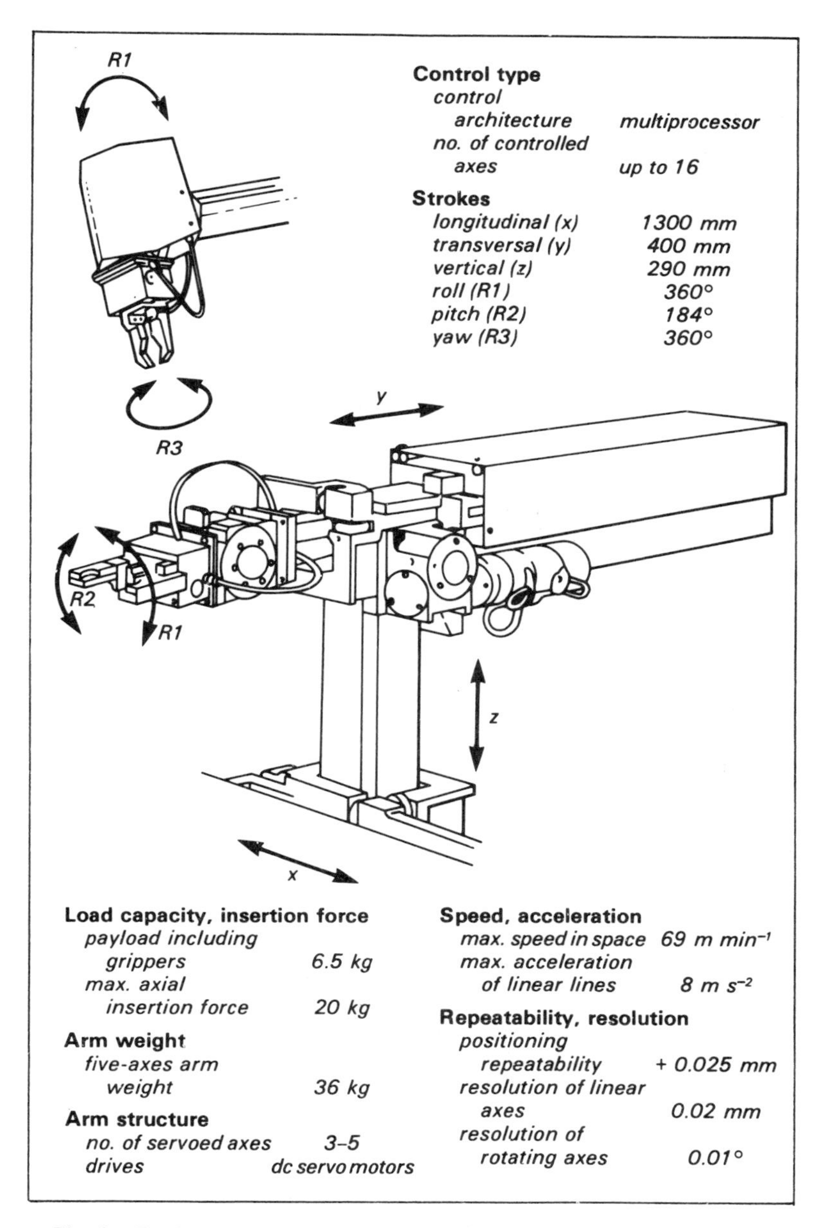


*Fig. 2 Technical specifications of the DEA Pragma A 3000 robot*

in the Chieri factory. (The DEA Pragma A3000 robot and technical specifications are shown in Figs. 1, 2.)

The success of this initial venture has prompted Aspera to subsequently purchase six other systems from DEA. Two systems are used for the assembly of valves, one for the assembly of con-rod piston and three for the subsequent stages that form the complete assembly.

## The product on which the robot is used

In order to give the DEA robot a realistic trial Aspera attempted to automate quite a complicated assembly task. In this way the company got a more realistic appraisal of the robot's performance. Taking all factors into account the final choice of application was a valve assembly mounted on a compressor motor (Fig. 3).

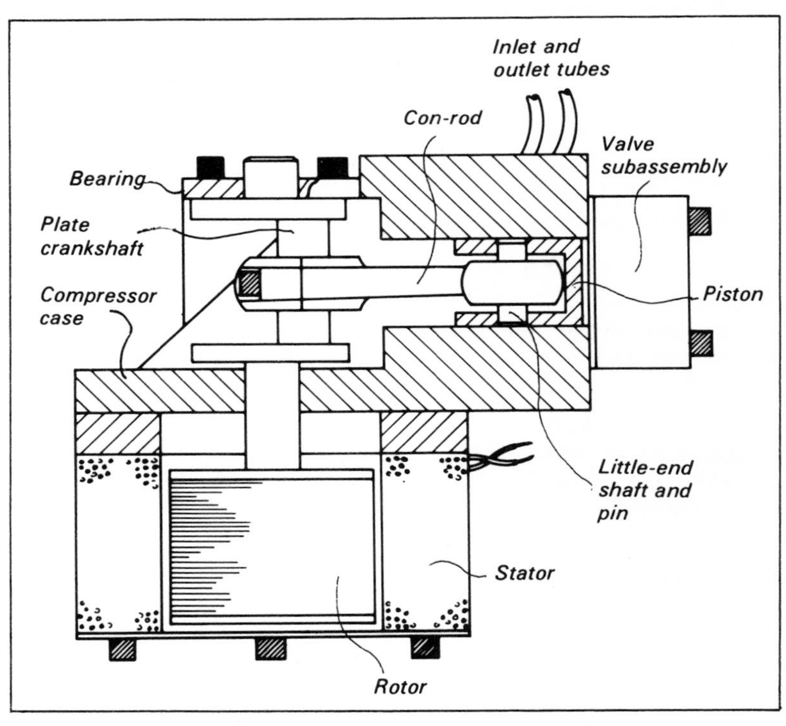


*Fig. 3 Cross-section of the valve plate assembly*

This relatively simple mechanism has some very precise tolerances, particularly associated with the piston and crankshaft. These tolerances require an accuracy beyond the specification of the majority of robots. As a result the part chosen for robot assembly was the valve sub assembly.

## Robot installation and workcycle

The layout of the first of the Aspera assembly robot installations is shown in Fig. 4. The operator has access to the hoppers numbered 5, 7 and 11 to load the valve plates, cylinder heads and screws that are used as a part of the assembly of the refrigerator valve. The operator is positioned well away from the two robot arms A and B; it is only when there is an emergency, or in the event of a feeder blockage, that there is any need for the operator to enter the safety cage. In such cases the robot arms are immobilised by a safety interlock on the entrance gate.

The robots have three basic linear axes of movement ($x, y, z$). In addition they can make three circular wrist movements (R1 $\pm$ 180°, R2 $\pm$ 92°, R3 $\pm$ 180°). This makes the robot ideally suited for the assembly task; the combination of the two arms creating an immensely flexible, highly accurate (0.02 mm and 0.02°) assembly machine. The robots are programmed using a point-to-point system that enables the robot arm movements to be coordinated in the most efficient manner.

The installation without the safety cage is shown in Fig. 5. The two robot arms are in the foreground, with the hoppers, jigs and fixtures in the background. The whole installation is firmly mounted on a solid, levelled metal table that reduces vibrations and enhances the accuracy of the overall system.

**Machine cycle description.** The operation of the DEA Pragma A3000 robots enables both arms to operate simultaneously, although they are interlocked to ensure that no crashes occur. Before examining the machine cycles it is as well to examine the compressor valve in detail (Fig. 6).

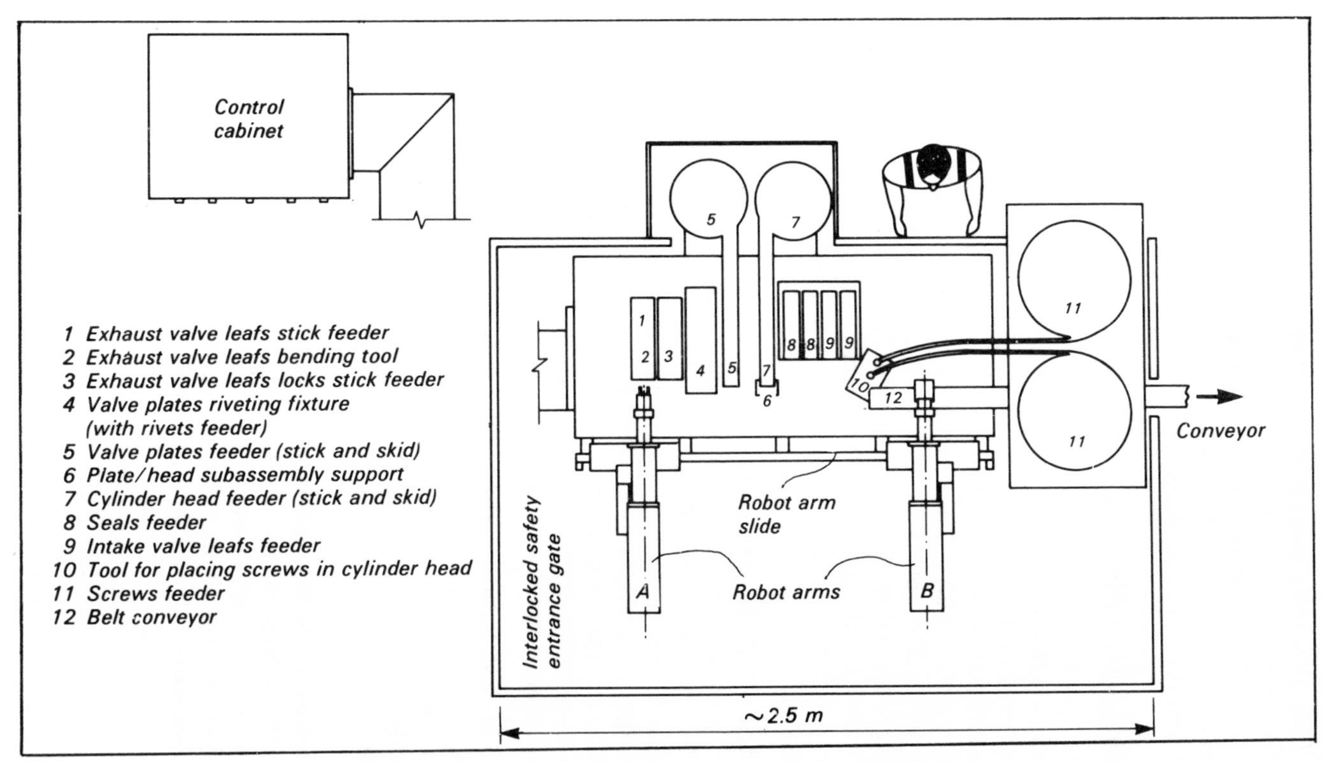


*Fig. 4 Layout of the Aspera assembly robot installation*

*Fig. 5 The actual installation (without safety cage)*

With reference to Fig. 4 the work sequence for arm A is:

(i) Valve plate preassembled in station 4.
- valve plate picked up from station 5
- valve plate deposited at station 4
- arm moves to station 2
- exhaust valve is picked up, bent and oriented
- valve leaf deposited on the valve plate in station 4
- arm moves to station 3
- leaf lock picked up
- leaf lock deposited on station 4

(ii) Riveting of station 4 is completed.

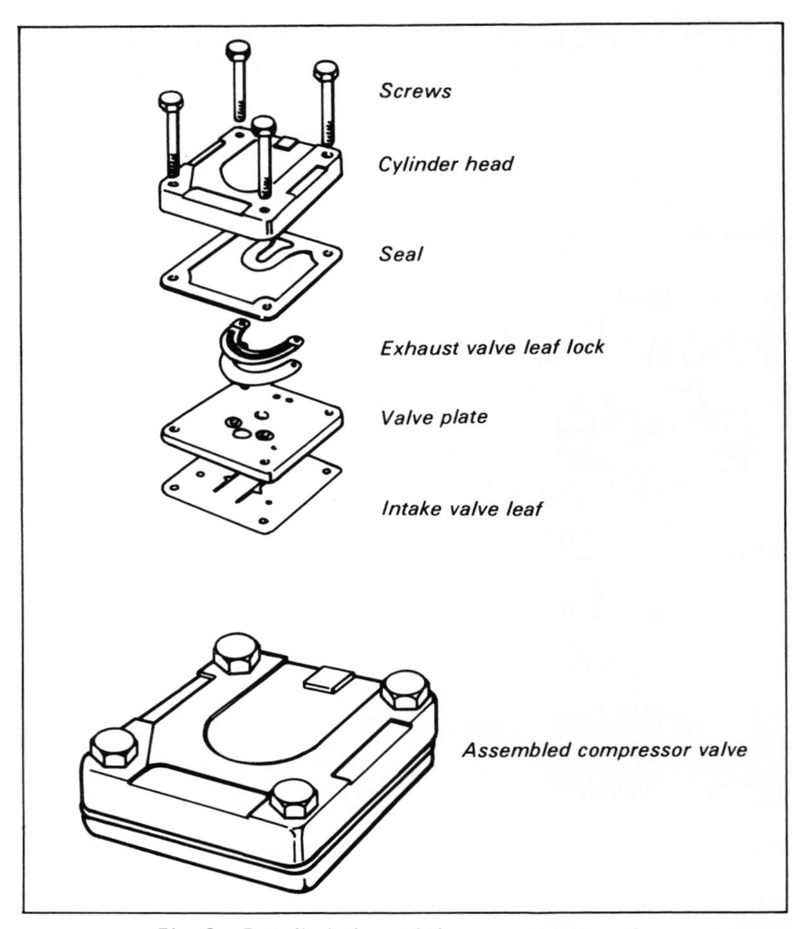


*Fig. 6 Detailed view of the compressor valve*

(iii) Preassembled valve plate is transferred to arm B operating area.
- — valve plate picked up from station 4
- — valve plate transferred to station 6 with a 180° wrist rotation

The work sequence for arm B is:

(i) Valve plate and cylinder head subassembly.
- — seal picked up at station 8
- — seal deposited on the cylinder head in station 6

— signal given to enable stage (iii) above to be completed
— arm moves to station 9
— intake valve leaf picked up
— intake valve leaf placed on valve plate in station 6

(ii) Valve plate and cylinder head subassembly transferred to station 10 for screws to be inserted.
— subassembly in station 6 picked up
— arm moves to station 10 with a wrist rotation of 180°
— reversed subassembly deposited in station 10

(iii) Screws are inserted in the subassembly in station 10.

(iv) The assembled compressor valve is transferred to the belt conveyor.

All these operations include an element of testing that is accomplished through force sensing at the wrist/gripper-end of the robot arm (Fig. 7). The sensors in the arm have an element of compliance with high and low trigger levels. If a component is missing, or if more than one is picked up, an error signal is passed to the robot controller. Depending upon where the robot is in the program, it will have some preset

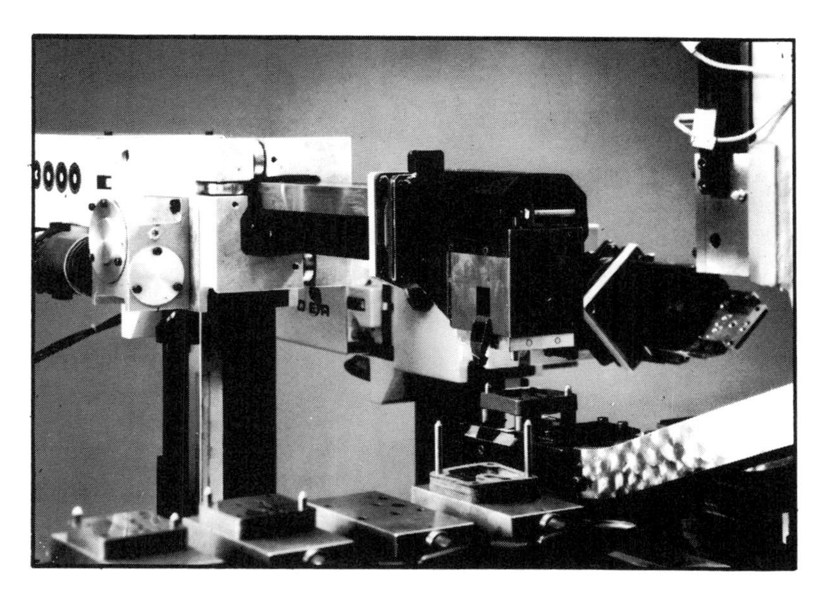

*Fig. 7 Testing is accomplished through force sensing at the gripper-end of the robot arm*

responses that will prompt some remedial action from the operator.

With this assembly technique the workcycle of the system is 10 s. The robot assembly cell requires the attention of a full-time operator, although three robot stations are serviced by one maintenance technician.

## Manual system

A plan of the old manual layout and a brief description of the manual work sequences associated with these workstations is shown in Fig. 8. Although the workcycle is about the same length for each workstation, the workcycle for the complete assembly is 30 s. Because of the length of the respective workcycle, stations 1 and 2 feed station 3. The output from the final station is loaded on to trays before being passed on to the main assembly tracks.

Both the robot system and the manual system are worked for two shifts a day. Each shift lasts eight hours with half-hour meal breaks.

## Designing, installing and running the system

Bearing in mind that this robot installation was a first for both robot supplier and user, the development time for the system was surprisingly short. In fact it took one year to design and build the system in DEA workshops and a further six months after installation to achieve full production. (Note, this is not usual when buying a robot system 'off the shelf'—in these circumstances 6–8 months would be a more typical period in which to achieve full production.)

To allow for the development of the system, it was designed to give the required production output when operating with only 80% up-time. In this way any improvement in performance would result in higher overall levels of output. Apart from some teething problems on the robot system itself the majority of problems were associated with the ancillary equipment (i.e. jigs, fixtures and feed mechanisms). Mixed in with these problems were those associated with component tolerances and Aspera has had to improve some component specifications for both the company and its suppliers.

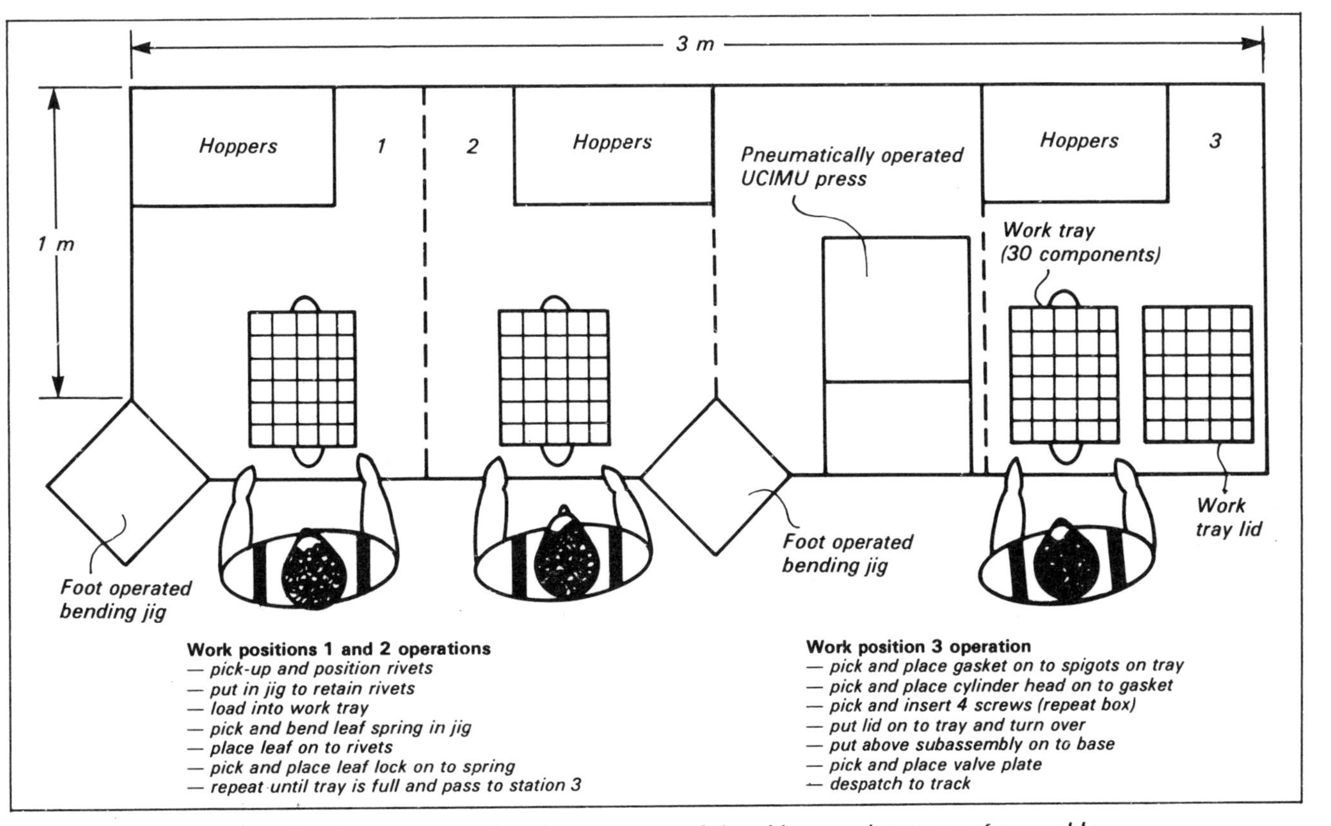


Fig. 8 Layout and work sequences of the old manual system of assembly

Typical of the problems that were encountered during this period were screws sticking in the feed lines from the hoppers (designated 11 in Fig. 4). This problem was all the more perplexing in that it only occurred after weekend or holiday shut-downs. The screws concerned were checked and found to be within tolerance, as were the feed mechanisms themselves. Eventually the problem was traced to the component hoppers which were topped up regularly from a fresh supply of components during normal production times. When components were left over a weekend or holiday the normal coating of oil on the screws drained away and this in turn led to the screws sticking in the feed mechanisms. The solution was simply to oil the screws after each subsequent break of more than two days. In itself this was not a very spectacular problem, however it does give an indication of the nature of the learning processes involved.

Another learning area is that associated with the maintenance of the robot system. Initially Aspera was very apprehensive about tampering with the robot system at all. The company was on a learning curve and in the initial phases independent of DEA support. The knowledge Aspera has accumulated has been put to good effect in the design of subsequent robot systems, and Aspera now achieves well over 80% utility from all systems.

## Labour relations

The Aspera workforce is unionised and the company has a relatively trouble-free association with the unions. The workforce is drawn from local villages and small towns. Few people bother to travel from Turin despite its close proximity. Even though the workpeople are from villages few have agricultural interests. In this way they vary from the workpeople in the south of Italy who frequently have smallholdings, and who use the factory as a means of supplementing poor incomes from the farms.

In spite of labour reductions (about 12 people) on the compressor assembly lines due directly to the introduction of robots, there has been little resistance to change. Aspera is looking to achieve further labour savings throughout its system with the introduction of more automation in the future.

## Financial considerations

The original complete assembly system cost about £100 000. The spares package for the system cost £10 000, with a 40% discount for buying it with the robot.

In general Aspera is looking for a payback on capital expenditure of about five years. With the assembly system payback is now expected to occur at between two and three years. The bulk of the savings associated with the system are generated from an increase in manufacturing efficiency, a decrease in direct labour costs and a reduction of work-in-progress.

## Important learning points

- When Aspera moved to automatic techniques of production from the conventional manual methods of production it had to learn to think of production in a completely new way.
- Where automation is to be introduced there has to be an element of compromise in production techniques and in product design.
- The production problems that have to be overcome become progressively more complicated the more you want to automate.
- Know what to expect of automation, in particular be aware of what it cannot do.

## The future

At present about 60% of Aspera compressor assemblies and about 10% of compressor parts tool-working are automated, and the company plans to extend automation throughout the manual production system. Beyond assembly operations Aspera is also considering the robotising of the arc-welding associated with the upper and lower cases that the compressor assemblies fit into.

**Acknowledgement**

The author thanks Dr. Lanfranco of Aspera, Sr. Bertolini of DEA, and Mr. Rollo and Mr. Badger of Fairey Automation (Swindon), for their assistance in compiling this study.

# *Case Study 2*
# Huard UCF—SCM
## (Chateaubriant, France)

HUARD's head office is situated in Chateaubriant midway between Rennes and Nantes in north-west France. Here Huard has three factories, two of which are dedicated to the painting and administrative facilities. Huard manufactures a wide range of agricultural equipment including ploughs, harrows, cultivators, sub-soilers, tillers, trailers and spare parts. The majority of products (85%) are sold under its own brand name although the company does undertake some subcontract work (15%) for other agricultural equipment manufacturers.

The company employs about 1700 people. At the main Chateaubriant plant 800 people are employed in the manufacturing departments and about 50 in the commercial and administrative departments. The forging factory employs about 200 people and the foundry about 450. (The remainder of the workforce are employed at a small factory at Carcassone near the Spanish border.) Situated in the heart of some of the richest farmland in France, many of Huard's workforce are people who have been displaced by the increasing use of machinery on farms. A small percentage of the workforce have small-holdings within a 20 km radius of the factory.

In Europe, the company's main competitors are Knevunelang of Denmark, although in France, Huard still dominate the market particularly in mould ploughs (52%) and reversible ploughs (70%).

## Company history

In 1850 Mr. Huard, a blacksmith in Chateaubriant, began to specialise in the manufacture of Braband ploughs and seed drills used for animal-based farming methods. This enterprise was successful and by the turn of the century he employed 20 people.

When Mr. Huard retired his two sons, Jules and Frances, took over the management of the company. The company reinvested its profits in its manufacturing facilities and by 1914 the company employed 140 people and sold over 5000 ploughs a year. This made Huard the leading manufacturer of ploughs in France. After the First World War large numbers of forge tools were installed and the animal-drawn farming equipment went into mass production. In 1929 Jules Huard bought the local foundry to supply his needs for cast iron.

During the Second World War part of the Huard factory was destroyed by aerial bombardment but was rebuilt on the same site. As France began to rebuild its economy, in 1948 Huard made the decision to begin mass production of tractor-powered two-way ploughs that were derived from the original Brabant ploughs. Other competitors were less skillful in anticipating the market trends and this enabled Huard to corner the market.

In 1957 Huard decided to complete the range of farming equipment that it offered to its customers. The company amalgamated with three other French companies to eventually form Huard as it is known today.

Jean Huard, the third generation of the family, is now President and Director General of the company.

Although Huard has consistently invested in capital equipment, until recently this has been of a more traditional nature. However, in 1976 the company began to investigate how it could use more automatic equipment. For arc-welding work there were two possibilities—fully automatic equipment or arc-welding robots. In the former the equipment that appeared most suitable was made by SAF. The choice of robots was not so clear since Huard had no experience of this technology. The company subsequently contacted the French Welding Institute who provided a list of possible robots.

In searching for a robot to suit its needs Huard began by analysing its product range which was considered to be suitable for robot welding. This analysis indicated that a robot which could weld long structures (3 or 4 metres) was required, but which only required limited depth and height movements. Field trials of both robotic and fully automatic equipment were undertaken.

The fully automatic equipment had only moderate success as it required constant adjustment and was perhaps too accurate for the weldments. The robots under trial were more successful. The smaller robots performed adequately but would have to be mounted on a trolley or rails to cope with Huard's length requirements. Among the larger robots tested was the Languepin robot which came supplied with a long *x*-axis as standard. This was supplemented with an extra axis—a programmable head- and tail-stock—which gave the robot almost ideal characteristics from Huard's point of view. With

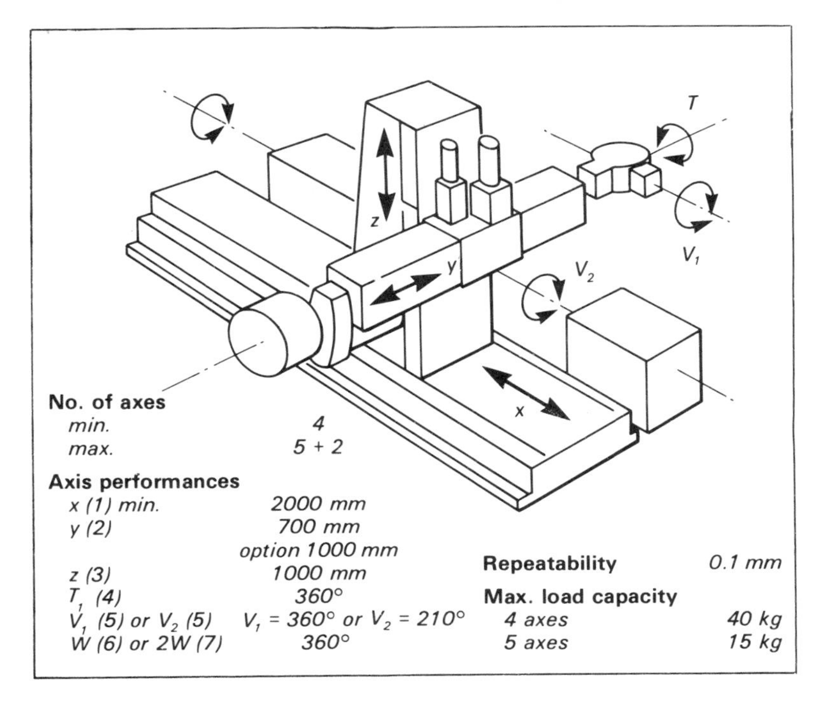


*Fig. 1 Technical features of the Languepin robot*

these characteristics and with successful field trials Huard finally chose the Languepin robot (Fig. 1).

## The products on which the robots are used

Although this case study relates to one specific application of its first robot, Huard uses the robot to arc weld many of the components that are used on its product range. One long running set of components that have been welded with robots are those associated with the many ploughs that Huard

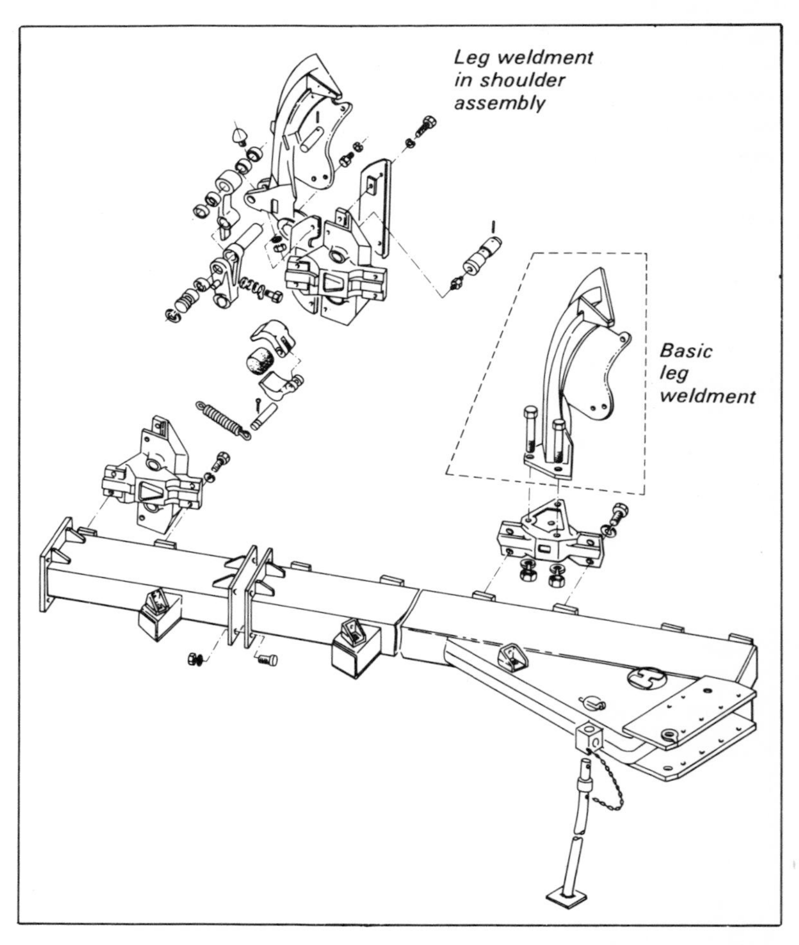


*Fig. 2 An exploded view of the main beam assembly of the diamond plough*

produces. Huard makes between 40 and 100 ploughs a day of varying types and sizes. The arc welding required from the robot system varies from simple 8 mm fillet welds to complicated profile welds requiring tapering fillets. (It is worthwhile stressing that Huard still does a lot of conventional manual welding in its welding department and employs 43 welders.)

The component is used in the manufacture of the Huard Reversible Mounted Diamond Plough. An exploded view of the main beam assembly of the plough is shown in Fig. 2. The basic structure on which the plough blades are mounted is a square section beam. This can be extended in length to carry extra plough blades. The blades are mounted both above and below the beam on shoulder assemblies. These are specially designed to allow the whole of the plough blade to retract in the case of a burried boulder being struck by the blade when being pulled by a tractor. Between the shoulder of the plough and the blade is the leg weldment.

The plough blade is quite a complicated assembly and is bolted on to the leg weldment (Fig. 3). The weight of this blade and the operational forces encountered in ploughing require the leg to be a substantial structure of complex profile and high mechanical integrity.

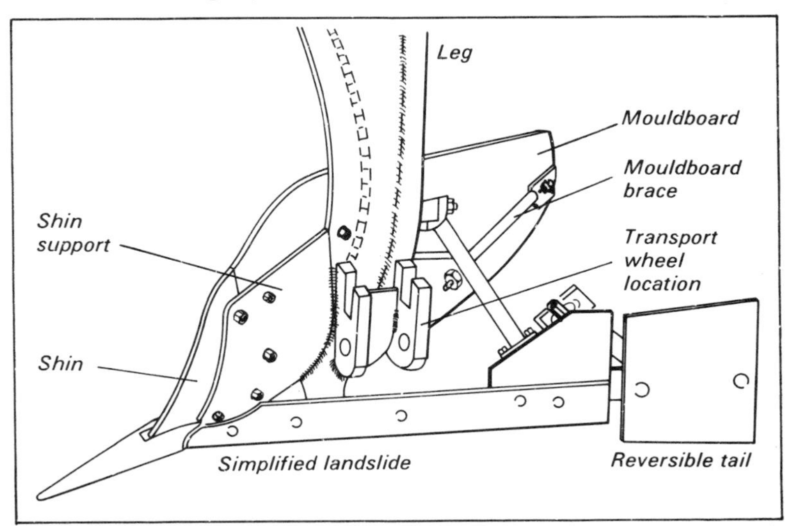


*Fig. 3 The plough blade assembly*

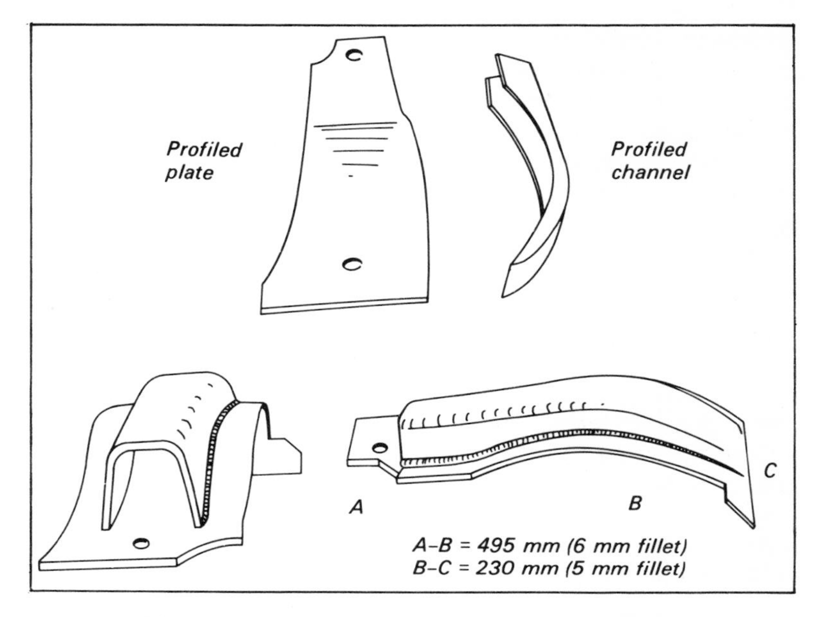


*Fig. 4 Components and assembly of plough leg*

Details of the leg weldment are shown in Fig. 4. It is made from two components, a profiled plate and a profiled channel. These are positioned together with curves matching and then clamped and welded. The weld is interesting in that it has a tapering profile that diminishes from 6 mm fillet between points A and B to 5 mm fillet between points B and C. This welding is done by the Languepin robot.

## Robot installation and workcycle

The general layout of the Huard robot arc-welding station is shown in Fig. 5. The installation is quite large occupying a floor area of about 10 × 5 m. The Languepin robot has a 6 m axis and can service two workstations. Normally the system works with one operator who would prepare the weldment in whichever workstation that is not being used by the robot.

The work is held in a jig mounted between a headstock and tailstock. The headstock can be driven to a position that is determined as a part of the robot's program. This enables the welding to be done in a gravity position giving better weld penetration and superior welding speeds. Power for the arc is

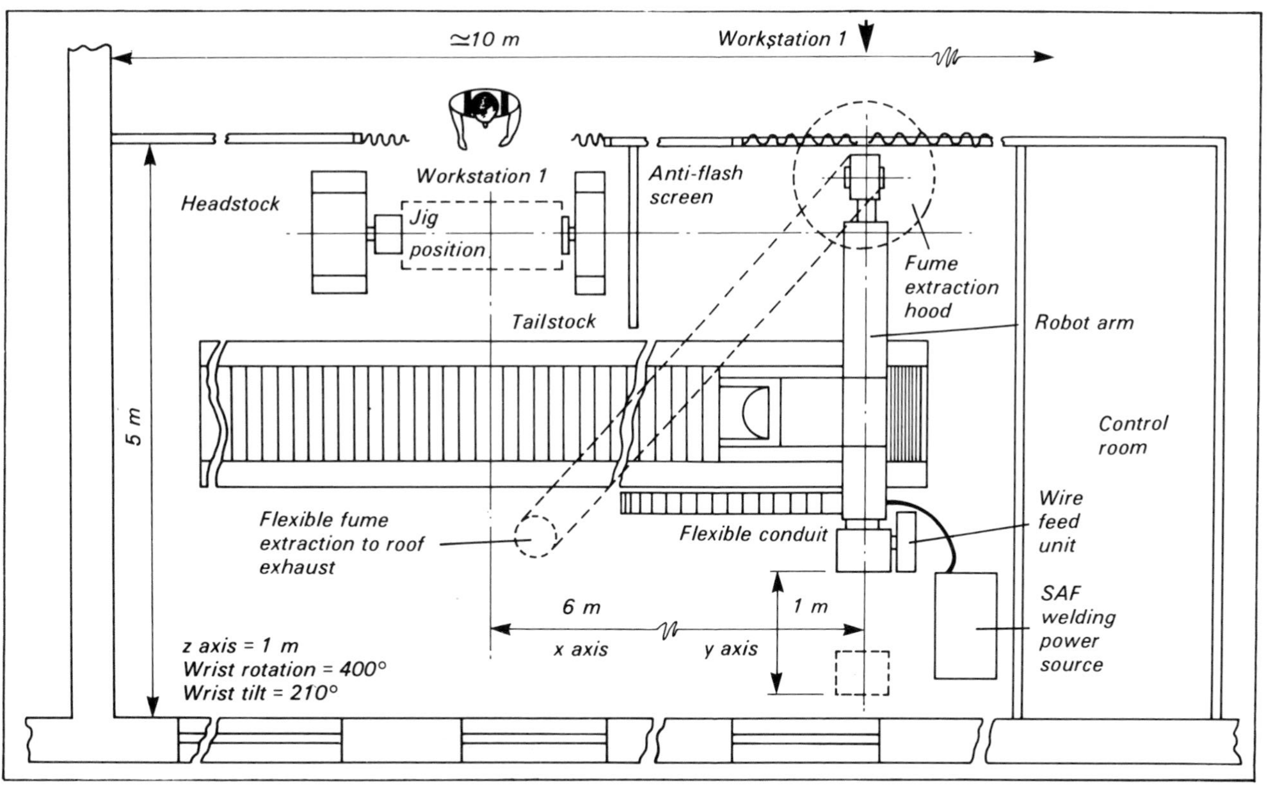


*Fig. 5 Layout of the Huard robot arc-welding station*

provided by an SAF welding set with a range of up to 340 A at up to 28 V. This welding set has a special interface that enables up to four preselected sets of current and voltage to be chosen by the robot controller without manual intervention. The wire feed for the welding set is positioned at the back of the $y$ axis of the robot and moves with the robot. The wire used for this application was 1.2 mm in diameter and copper coated. Fume extraction for the welding is provided by a hood mounted above the welding head. This hood is connected via a flexible hose to a fume extraction unit in the roof (Fig. 6).

The robot in this installation has five axes. Three linear axes of 6 m, 1 m and 1 m, and two circular axes at the wrist of 400° and 210°. The robot has a positional accuracy of ± 0.5 mm. The headstock is in effect the robot's sixth axis.

Other than to change the wire reels the only reason the operator has to enter the robot compound is to load and unload the welding jigs. Because of this there are no special safety precautions inside the robot compound. Normally only trained operators or maintenance technicians are allowed in the space between the robot and the wall. Finally, the welding torch is despattered automatically every five to six cycles. The inside of

*Fig. 6 Fume extraction for welding—the hood is connected via a flexible base to a fume extraction unit in the roof*

the torch shroud is scraped with an internal blade and then sprayed with an anti-spatter compound. This has proved to be an effective way of coping with this problem.

**Manufacturing cycle description.** The basic manufacturing cycle for this component is quite simple and may be summarised as follows:

(i) The profiled plate is placed in the welding jig.
(ii) The profiled channel is positioned and clamped over the plate.
(iii) The operator signals to the robot control system that the loading is complete.
(iv) The robot moves to the freshly loaded work position when the previous cycle is completed.
(v) The welding jig rotates to present the first weld in the vertical (or gravity) position.
(vi) The robot welds the seam and moves clear when complete.
(vii) The welding jig is rotated to present the second weld in the correct position (Fig. 7).
(viii) The robot welds the seam and moves clear of the jig.
(xi) The robot moves to the other work position to continue the workcycle from stage (iv) above. The jig is simultaneously rotated into its original position for unloading.
(x) While processes (iv) to (ix) are being completed the operator unloads and reloads the work jig.

(Note, the actual welding path and robot movements are carefully planned by Huard's workstudy department using wooden models of components to ensure accuracy.)

Using this workcycle it has been possible to achieve a workcycle time of 13 min 28 s with a duty cycle of 80% (i.e. arc time : non-arc time). This compares with a manual workcycle time of 46 min 24 s. An operator loading the welding jig is shown in Fig. 8.

One operator services two workstations and one maintenance technician will service three robot installations. The operators are not fully skilled welders. The system described produces less than 1% rejects and these usually only require minor corrective action. Normally the robot is operated for two full shifts a day (17 hours).

*Fig. 7 The welding jig is rotated to present the second weld in the correct position*

*Fig. 8 An operator loading the welding jig*

## Manual system

The layout of Huard's manual welding booth is shown in Fig. 9. This is a typical layout and was used for any components manufactured by the company.

In this system the various components that make the final weldment are stowed in the stillage marked A. The welder collects the components from these storage units and fixes them in the jig on the welding table F. The various components are then welded before finally being transported, again by the operator, to stillage B. When the batch is completed stillage B is removed by a forklift truck and replaced by an empty stillage. With this system, a great deal of operator movement is required and this is naturally inefficient. The manual system operates on a two-shift system with half-hour meal breaks.

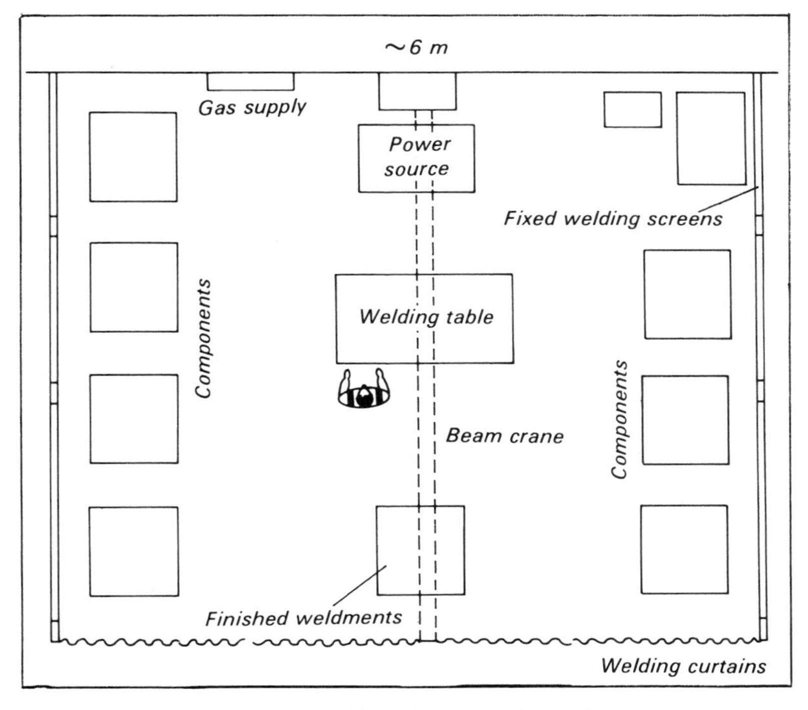


*Fig. 9 Layout of Huard's manual welding booth*

## Designing, installing and running the system

By 1980 Huard had devoted a lot of time to evaluating its needs and the robots that were available. Conceptually the system had to be simple and flexible. The robot system needed to have a long *x* axis to accommodate the nature of their products. However, Huard was always aware that it would have to design and build its own jigs and fixtures and did not expect any robot manufacturer to undertake this task.

The choice of the Languepin robot was obvious given the operational requirements of the system. Indeed the robot was almost unique in that the 6 m axis was offered as a near standard feature. Once this choice had been made it was important to get the robot system on to the shop floor as quickly as possible. An area had been prepared away from the existing welding shop. Huard decided that it would take a shorter *x* axis than desired to get a quick delivery—with the extended arm being delivered within two months. Languepin delivered the robot on a Friday afternoon and by the following Wednesday morning it was working on a production task.

These early days were important as they enabled Huard to learn about using the robot system and about the design of welding jigs and fixtures. Also, minor technical problems with the robot system were eliminated. Operator training was very important since it was Huard's intention to use unskilled operators. Maintenance training was also important, although Huard declare that it's most important diagnostic tool is the telephone outside the control booth!

Once into production the robot system has been relatively trouble free. With regular maintenance Huard expects about 95–98% up-time for the system. The work capacity of the robot translates into higher productivity which initially ran to about five times that of the conventional manual system. However, there has been an increase in manual production since the introduction of the robot, and now the ratio is about three times that of the manual system.

## Labour relations

Huard is unionised and the workforce belongs to either the socialist CFDT or the communist CGT; these unions were involved in negotiations prior to the introduction of the first robot.

Although not directly related to the introduction of robot technology, Huard had noticed that over a period of time the forging and welding activities of the company attracted more than normal levels of labour difficulties. Eventually the forging activities of the company were moved to a green-field site in the Chateaubriant vicinity. This meant that people worked in a very pleasant environment, and has subsequently resulted in lower levels of labour dissatisfaction. Huard could not contemplate a similar move of its welding activities, however it did hope that eventually robots would help with some of the more arduous welding tasks.

In spite of the company's genuine desire to introduce better working conditions with the robots, the trade unions were suspicious of management's intentions. After negotiations an agreement was eventually reached based on managment's willingness to bring in work that was being subcontracted at the time. Also the company gave undertakings on redundancy and redeployment. It was also agreed that while unskilled workers would load and unload the robot jigs, skilled workers and technicians would teach and maintain the system.

Since the introduction of the first robot system and its acceptance by the workforce, another system has been installed and yet another planned. Evidence shows that despite initial fears the workforce does enjoy working with the new equipment.

## Financial considerations

The original robot welding system was purchased in 1980 for about £65 000. All jigging and fixtures were manufactured by Huard and no figures were available for this expenditure.

Initially Huard was a little unsure of the expected payback of the system. A conservative estimate of 3–4 years was made. In reality the company will achieve a payback on the first system of two years—one of the factors that has encouraged Huard to purchase further systems.

The two-year payback period can be attributed to the threefold increase in productivity of the robot system, giving improved labour utility, lower direct labour costs per unit produced, higher quality, and lower number of rejects and reworks.

## Important learning points

- Huard has found that it had to rethink its normal manufacturing methods to ensure that work systems are correct first time. This usually means that more thought has to go into the production engineering of the system.
- Workflow through the robot welding section is faster because there are fewer ancillary operations and work movements.
- The work of the robot systems is best controlled by one person whose sole responsibility is to optimise the system performance and to organise the day-to-day running of the system.
- Design the robot system for productivity not sophistication. The more sophisticated the system is, the higher the probability of breakdowns.
- Now that Huard is familiar with the characteristics of the robot system it can prebend some components to accommodate welding distortion. In turn this gives the opportunity to simplify some of the welding jigs and fixtures.
- Robots have reduced the work-in-progress in the sections in which they are used by 10–20%. Also, inventory costs have been similarly affected.
- Huard can now more easily balance its production to help in the supply of the seasonal demand for certain products.
- The robots should be kept working.

## The future

As previously mentioned Huard is already considering using more robots in other welding and non-welding areas of manufacture. These include: painting, palletising and loading.

The company is also investigating the possibility of using a robot-fed machining centre in its workshop.

**Acknowledgement**
The author gratefully acknowledges the help of Mr. Provost of Huard, and of Mr. Lesbleiz and Mr. Dalton in compiling this case study.

# *Case Study 3*
# Imhof Bedco Special Products Ltd.
## (Harpenden, UK)

IMHOF BEDCO is part of the Electronic Enclosures Division of Phicom plc. (Phicom is a holding company who has interests throughout the light engineering and electronics industries in the UK and Europe.) Imhof Bedco is located at Harpenden, about 20 miles north of London.

The company manufactures sheet-metal enclosures, panels and frameworks to customers' specifications. Although the company is active in many market sectors, its main business is derived from the electronic, computer and vending machine markets, as well as undertaking a substantial amount of government work. However, as is common with many companies in the contracting business, the market mix is quite volatile and can change with surprising rapidity.

Imhof Bedco employs about 200 people on its two Harpenden sites. These sites are located within about 300 m of one another. Historically, the split was due to the growth of the company and the unavailability of a large single site in the vicinity. The main factory houses the administration and fabrication facilities, while on the other site the company finish and pack their products. The majority of the workforce are skilled and semi-skilled working with a large variety of conventional manual equipment, as well as numerical control (NC), computer numerical control (CNC), direct numerical control (DNC) and robot equipment.

### Company history

Imhof Bedco has a long and interesting history. As might be imagined its name came from the amalgamation of two companies.

The Imhof side of the company was originally founded by Daniel Imhof in 1845. From his shop in New Oxford Street in London he sold musical instruments and his world famous Orchestrion. In 1896 Daniel Imhof's son, Alfred, took over the running of the company. Alfred had inherited his fathers interest in new inventions and introduced many new products associated with the rapidly growing musical reproduction market. The company developed the first seamless gramaphone horn (made famous by the HMV dog!), bamboo gramaphone needles to reduce record wear, and the first public address system driven by compressed air.

In 1918 Alfred Imhof's daughter took control of the company and she continued with the company's close association with entertainment technology. The company was in the vanguard of this technology and was among the first to offer radiograms, televisions and hi-fidelity equipment. Eventually the company diversified into standardised sheet metal enclosures. As the years passed so the manufacturing side of the business grew and eventually separated from the retailing outlets.

It was in 1935 that Sir Edward Beddington Behrens founded Bedco. Sir Edward, a keen horticulturalist, began by manufacturing greenhouse heaters and propagators. As time progressed so Bedco diversified into the production of office furniture and sheet metal cabinets. When Sir Edward died in the late 1960s his company was taken over by Imhof. In 1972 the company was bought by Plantation Holdings who have subsequently become Phicom plc.

Throughout the history of both companies there has been a willingness to accept and apply new technologies. This is reflected in the attitudes of the existing management who in 1979 decided to embark on a corporate strategy of growth based upon the concept of high manufacturing efficiency.

A review of the manufacturing facilities at Harpenden indicated that if significant growth were to be achieved then some radical thinking about the company's manufacturing technology would be necessary. A great deal of its equipment was relatively old and labour intensive. Using similar techniques could only lead to growth by increasing the direct labour force and this would prove impossible in the existing factory space.

This led the management to the conclusion that growth must come from higher productivity based on capital intensive production and automation. The main production bottlenecks in the system were at the front- and back-ends. Naturally these were the first areas to be considered for improvements.

In 1980 the company's investment programme began at the front-end of the system with the purchase of a new CNC turret press to replace four manual presses. This was augmented by a new NC brake press and the retro-fitting of NC controllers to all hand-fed brake presses. This investment greatly enhanced the performance of the company's forming shop. The subsequent improvement in quality and efficiency has more than justified the decision.

Unfortunately the problems associated with the finishing shop were not quite as straightforward. The root of the finishing problem was the lack of paint-spraying capacity. Simple solutions to this problem such as paint reciprocators or paint dipping were not adequate because of the wide range of complex components that required many differing paint finishes.

It was only after lengthy discussion with paint experts and robot consultants that the company realised that it would have to use robots to solve its problems. Subsequent analysis showed that the required system would have to operate under some form of numerical control.

The final specification for the robot paint-spraying system called for three robots (Binks Bullows RAMP, Figs. 1, 2), three paint-spraying booths, a conveyor, and a 48 m double-pass stoving oven. A DNC computer would control the whole system. This system was ordered and subsequently installed in two phases. In July 1981 the basic system was implemented and in May 1982 the DNC controlling system was commissioned.

## The process in which the robot is used

Imhof Bedco does not make a product as such; instead the company offers a range of services to customers on a contract basis. These services include designing, prototyping, tooling, cutting, piercing, blanking, forming, welding, preparation, finishing, assembly and packing.

*Fig. 1 The Binks Bullows RAMP robot*

Within the finishing shop that houses the robot paint-spraying system there are five processes. These are (in sequence) degreasing, phosphating (or chromating), preparation, painting and inspection. All components in the processes of manufacture acquire a layer of oil and grease. (Some of these contaminants are already on the steel when it is delivered to the factory.) If these were left on the metal it would prevent correct paint adhesion and give a poor finish to the product.

Consequently components are either hot water and detergent scrubbed or Genkleen dipped. When clean and dry the components are phosphated. Phosphating is a process that chemically roughens the surface of the metal to ensure good paint adhesion. Once phosphated, components have any minute scratches or surface aberrations removed by either filing or grinding-off. This is necessary to achieve the high quality painted finish required by customers.

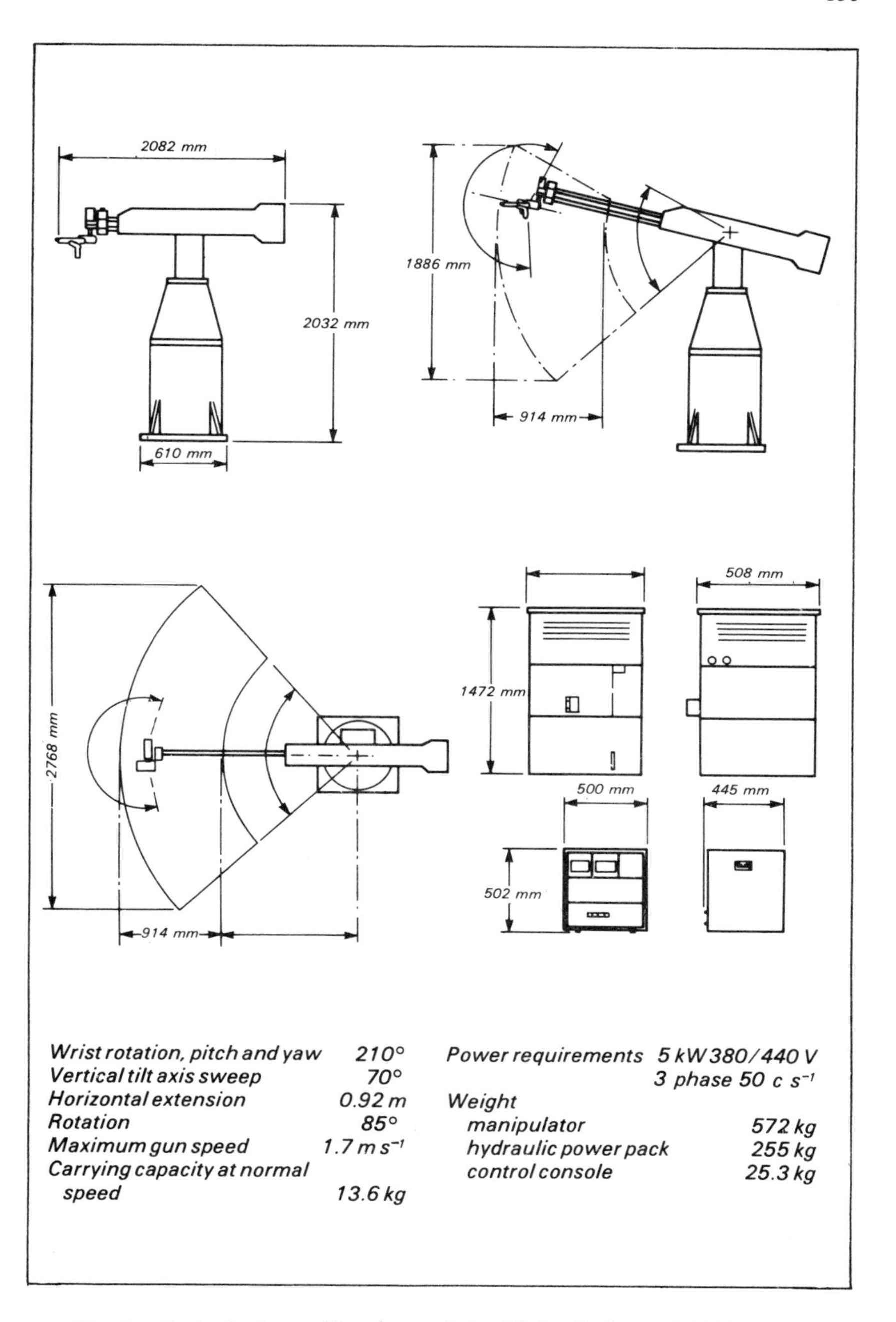


*Fig. 2 Technical specifications of the Binks Bullows RAMP robot*

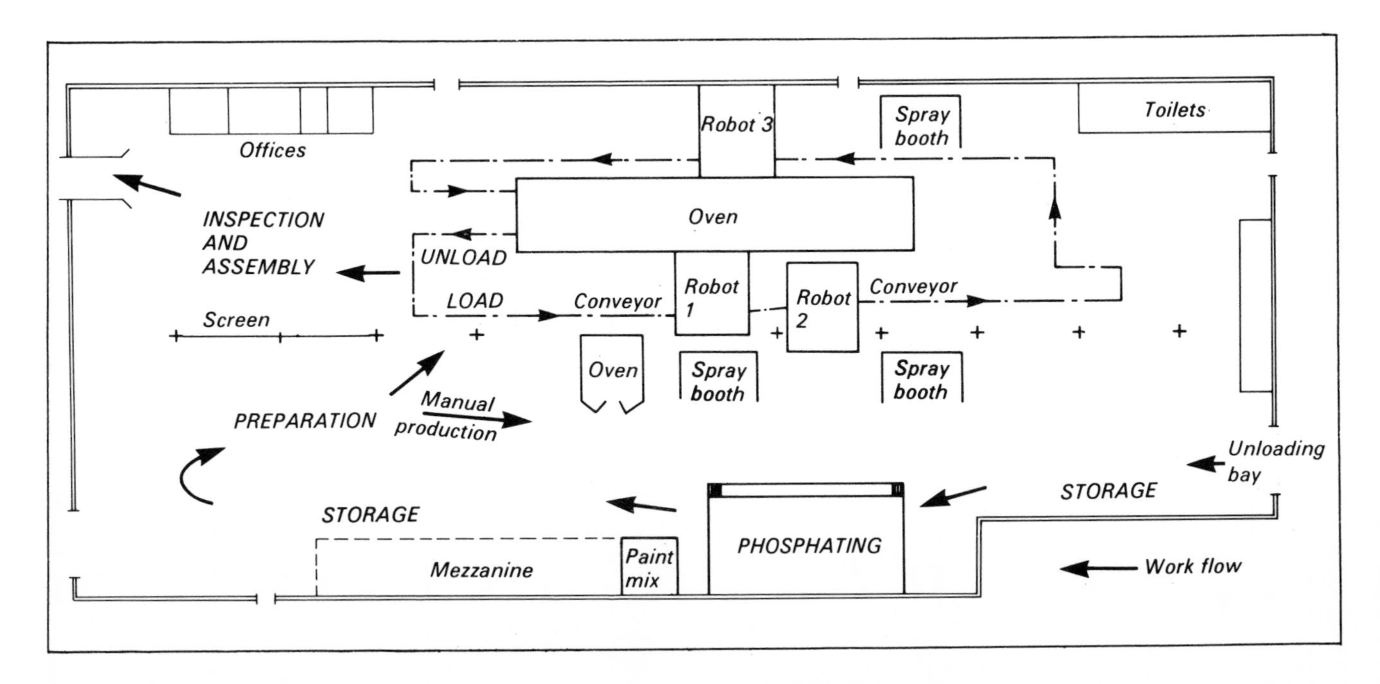


Fig. 3 Layout of the robot paint-spraying shop

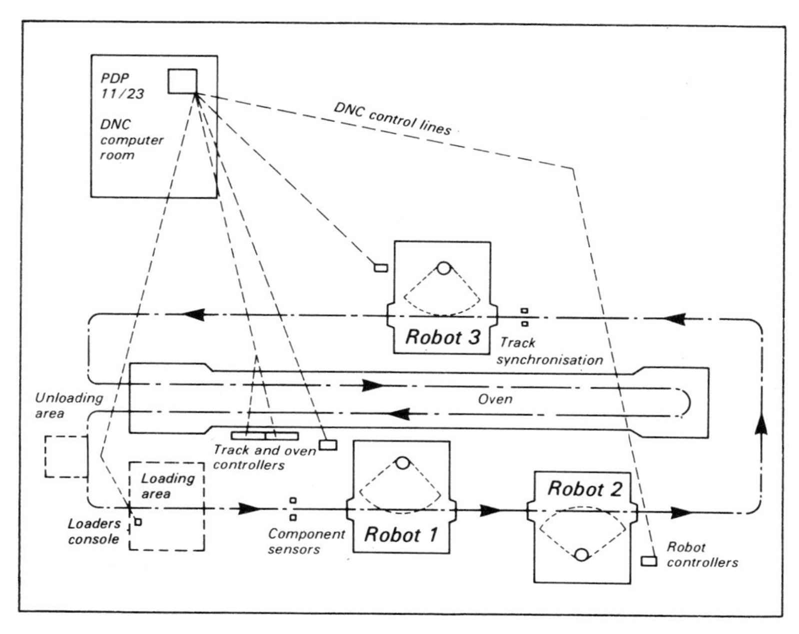


*Fig. 4 Layout of the DNC robot paint-spraying shop*

Components are then passed to either the robot line for painting, or to the manual spraying booths. (The manual facility is retained for awkward components or very small batches.) Once painted and stoved, components are inspected before being passed to assembly, packing and despatch. The layout of the paint-spraying shop is shown in Fig. 3.

## Robot installation and workcycle

The basic arrangement of the various system components in the robot spraying track is shown in Fig. 4. The system can work in two basic control modes—semi-automatic and fully automatic. For simplicity's sake, the semi-automatic mode of operation will be considered first.

The task of the robot system is to spray components that are hung on the conveyor (Fig. 5) to acceptable customer standards. To complete this task three robots are required. Robots 1 and 2 spray the respective sides of the components that face them (or which can be reached within the normal articulation of the robot).

*Fig. 5 Components are hung on a conveyor throughout the paint-spraying operation*

Once sprayed, components circulate in free air to allow solvent to evaporate or 'flash off.' Some components require a texture coating on one side and this is applied by robot 3 on top of the base coat before stoving. The oven temperature and track speed are adjusted to give the desired finish and to suit the paint being used. Throughout the process the robots are all synchronised to the conveyor speed. Sensors tell the robots when and when not to spray.

When in this mode of operation, the DNC computer has no effect on the track at all. The robots in the system are programmed off-line using a manual lead-through system, whereby the robots are taught by the operator who physically guides the robot through the desired movements. This program can then be either replayed or stored for future use.

The system components are Binks-Bullows RAMP robots (see Figs. 1, 2), Modan no-pump waterwash spray booths, a Solvex-Marshall conveyor, a Gee and Gee double-pass air-recirculating oven, and Sammes electrostatic spray-gun systems.

The DNC computer control system effectively sits above the semi-automatic system and when switched on takes control and integrates its performance. So, for instance, when a new job is to be put on to the robot track, the DNC computer will automatically set the track speed and oven temperature. Then it will load the previously stored programs into the three robot controllers and finally will inform the loader to begin loading components on to the track.

The paint changeovers and equipment cleaning between batches are done manually. These people pay particular attention to the quality of the painting and adjust the painting conditions if necessary. When under full DNC control the robot line's output is about three times faster.

The major visible difference between manual control of the robot paint-spraying system and the DNC operation lies in the nature of the job records that are kept. In the manual control system the records tend to be minimal and largely concerned with the paint colour and special requirements. With the DNC system the following information is recorded for each component:

—batch number
—order number
—part number
—repeat number
—conveyor speed
—oven temperature
—robot to be used
—paint type
—paint colour
—air pressure
—spray-gun type
—electrostatic voltage
—resistivity
—pot pressure
—solvent ratio
—loading arrangements
—paint programs for each robot

All information with the exception of the painting programs are usually entered into the DNC program by the shift supervisor when the job is first put into the system. The painting program for the robots is taken either at the beginning or end of each batch when the programs are first taught by the operators. Alternatively the programs can be stored in the robot controller memory and retrieved at the end of each working day.

When in normal operation the daily schedule for the robots is provided by the company's production control department. These schedules are fed into the DNC computer by the supervisor. This schedule is then printed by the computer and issued to the foremen and operators on the shop floor. (A sample DNC printout is shown in Fig. 6).

```
TODAYS SCHEDULE ON 13-JUN-82 AT 14:36:00

BATCH NO:      1 ***********************************
ORDER NO: 7663      PART NO: PT1      MAXI COVER      REPT NO: 0
QUANTITY: 120      SPEED: 7.0      OVEN TEMP: 80
CUSTOMER:          PART: MAXI COVER
HANGING ARRANGEMENT: DIAG. 33

REMARKS:
USE 10 INCH AND 8 INCH HOOKS

ROBOT 1

PAINT TYPE: EPOXY                COLOUR: BLUE
VISCOSITY: 30 SECS               PAINT NO: 11112233
GUN TYPE: ELECTROSTATIC          ELECTRO. VOLTAGE: 40 KV
RESISTIVITY: .0088               MAT/POT PRESSURE: 40 LBS
POT LIFE: 20 MINS                SOLVENT NO: 223344
SOLVENT RATIO: 1 TO 4            AIR PRESSURE: 100 LBS

REMARKS:

ROBOT 2

PAINT TYPE: EPOXY                COLOUR: BLUE
VISCOSITY: 30 SECS               PAINT NO: 11112233
GUN TYPE: ELECTROSTATIC          ELECTRO. VOLTAGE: 40 KV
RESISTIVITY: .0088               MAT/POT PRESSURE: 40 LBS
POT LIFE: 20 MINS                SOLVENT NO: 223344
SOLVENT RATIO: 1 TO 4            AIR PRESSURE: 100 LBS

REMARKS:

ROBOT 3

PAINT TYPE: EPOXY                COLOUR: BLUE
VISCOSITY: 100                   PAINT NO: 334455
GUN TYPE: MANUAL                 ELECTRO. VOLTAGE:
RESISTIVITY:                     MAT/POT PRESSURE: PUMP
POT LIFE: 4HRS                   SOLVENT NO:
SOLVENT RATIO:                   AIR PRESSURE:

REMARKS:
SPATTER FINISH TO BE CONSTANT ALL OVER PANEL
```

*Fig. 6 Sample DNC printout*

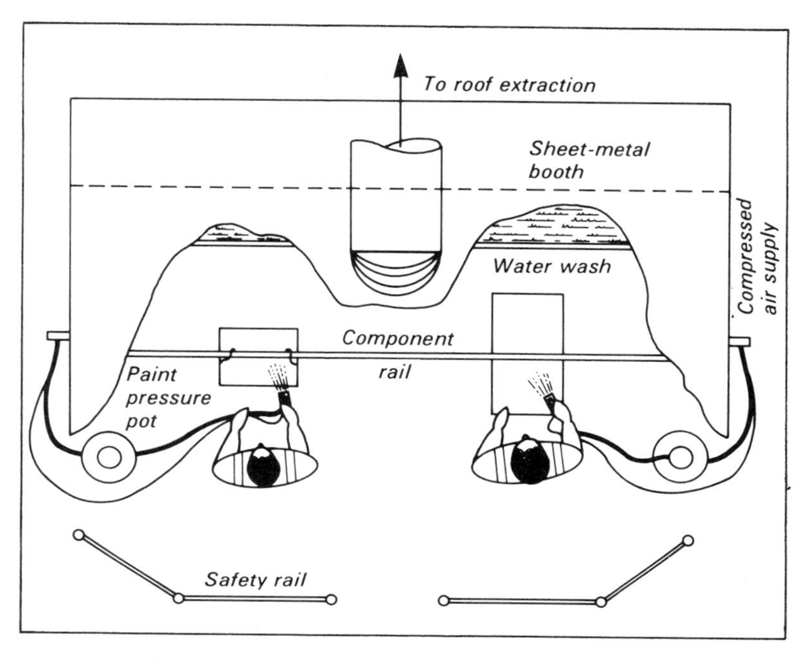


*Fig. 7 Layout of the old manual paint-spraying booth*

## The old manual system

Manual paint spraying has changed little in nature over the years. At Imhof Bedco it is possible to see the old manual system in parallel operation with the robot system. The company has retained it as a standby in case of robot breakdowns and partly for previously mentioned reasons.

The layout of a typical manual booth is shown in Fig. 7. Paint and solvent are mixed to the required viscosity and loaded into the paint pressure pots. The pressure in these pots is raised and this forces paint through to the spray gun. Compressed air is also fed to the spray gun itself. When the trigger is pressed a mixture of air and paint is ejected in a mist and directed by the operator on to the component.

Once sprayed the components are hung on racks for transportation to the stoving oven, after which the components follow the same process as the components sprayed on the robot line.

## Designing, installing and running the system

The original system design was carried out by the company's production managers and consultants. The robot system under manual control was installed to schedule and cost during the annual works holiday. This meant that the system was ready for production testing when the factory re-opened in August.

The early months of operation of the system were devoted to learning how to use the robot system efficiently and effectively. Many technical problems were encountered with the robots and took a long time to rectify with such a complex system. Other problems associated with the painting process were harder to overcome, in particular the robot spraying of texture coats. Eventually this problem was traced to the thixotropic nature of the paint, and an alternative to the pressure-pot system of paint feeding was designed.

These problems were costly in terms of lost production and effort. This inevitably led to the delay of the introduction of the DNC system. Many of these problems could have been avoided if some aspects of the system development had been performed off-site rather than on-site.

Learning from the installation of the manual control robot system the management was more cautious with the DNC system. The system was developed and proved in principle at the system designers. Then each part of the system was tested cautiously out of working hours on-site. The final commissioning of the system took place over a bank holiday weekend. This cautious approach paid dividends and the DNC system worked smoothly almost from the start.

With as complicated a system as this it takes longer to optimise the performance than with simpler systems. The reason for this, as Imhof's management believe, lies in the need for organisational adaptation. In particular, the DNC operation demanded more from the management control system than did the previous system. However to counter-balance some of the problems, there has been an increase in the quality and consistency of the work produced.

## Labour relations

Throughout the period prior to the introduction of the system and after management had decided to adopt the new technology, the workforce was kept informed of management intentions for the paint shop. The trades union (AUEW, Amalgamated Union of Engineering Workers) was consulted and agreements reached; a 'no redundancy' policy was adopted by the company.

Once the system was installed the management instituted a programme of continuous operator and management training. In addition, newsletters were distributed to the workforce throughout the factories to keep them up-to-date with the company's progress. These efforts have kept the attitudes of the workforce very positive, and this is critical for the future.

## Financial considerations

The reason for introducing robots was to achieve growth, and therefore payback has not been a prime consideration in the company's judgement of the system. However, the company expects a payback period of between two and four years on the £300 000 investment (made over a period of 18 months). The basis for this estimate is:

(a) greater productivity,
(b) lower rejects,
(c) higher consistency of finish,
(d) lower paint usage,
(e) savings in power used.

Having identified these savings the company stresses that payback calculations are difficult and to some extent will be dependent upon future investments in other parts of the manufacturing system.

## Important learning points

- Imhof Bedco has found that in order to maximise the utility of the investment it had to tighten its management control systems.
- The absolute minimum of on-site development should be undertaken when robot systems have to go straight into production.

- The competence of people selected as subcontract agents should be carefully monitored. In the long run this can be as important a factor as the goods offered for sale.
- Training and encouraging the right attitude among the workforce is vital to the on-going success of the system.
- The correct specification of the system as a whole is as important as the specification of the individual parts of the system.

## Future plans

Right from the outset of Imhof Bedco's move into automation it realised that its commitment was to a complete computer-controlled manufacturing system. This case study only describes one of the phases that the company is undertaking.

**Acknowledgement**
The author thanks Mr. Glossop of Imhof Bedco for his assistance in compiling this case study.

# *Case Study 4*
# Ramnas Bruk (Bulten-Kanthal AB)
## (Ramnas, Sweden)

THE RAMNAS BRUK factories are situated at Ramnas some 20 km to the north of Vasteras. One factory makes chains of varying sizes, while the other makes stainless steel kitchen and washroom sinks and stainless steel bowls. This case study is based on the robot installation in the second of these factories.

Ramnas Bruk is a subsidiary of Bulten-Kanthal AB whose head office is in Hallstahammar near Vasteras. Bulten Kanthal has twelve subsidiary companies in Sweden, and seventeen other companies trading throughout the world under the Bulten or Kanthal names.

Ramnas Bruk employs about 100 people at its Ramnas site—about half of them working in the stainless steel plant. Nearly all the workforce live close to the factory which is situated deep in the Swedish countryside. Many have worked in the factory for several generations.

Output from the factory goes via road or rail to Vasteras; the rail link being particularly important in the winter when the roads are often impassable. About 60 000–100 000 sinks and 200 000 bowls are produced each year.

The company's main competitors in Sweden are IFO and Gustavsberg. Even though Ramnas Bruk is the smallest manufacturer of stainless steel kitchenware, it has 40% of the Swedish home-market and exports about 10% of its total production. On the Ramnas site the turnover is about 30 million Swedish Kroner (£3 million) per annum.

In recent years the company has had rather a difficult time, largely due to the rapid decline in the building of new houses in

Sweden. This of course is a reflection of the depression that has prevailed throughout 1981–82.

## Company history

Ramnas Bruk can trace its history back to 1590 when a Royal forge known as Kungshammaren (Kings Hammer) was established in Ramnas. This forge manufactured wrought iron from pig iron. In these early days and throughout the 17th and 18th centuries the factory was the focus of the local life. Payment of wages was often partly in kind and many of the company's original records show payments to workpeople of grain, fish, salt, tobacco, etc.

In the mid 18th century a canal was built from Smedjebaken to Lake Malaren and this provided a great fillip to the small manufacturing communities that lay on its course. Prior to the canal all products had to be drawn by horse to the markets and distribution facilities of Vasteras.

At the beginning of the 18th century the Ramnas factory was bought by Magnus Schenstrom of Vasteras and Jacob Tersmeden, a local alderman. The factory remained in the hands of these families until 1830 when Colonel P.R. Tersmeden took sole ownership. In 1943 it was bought by Bultsfabrik AB.

The manufacture of wrought iron continued for the early days until the Franco-Prussian war of 1870–71. This war had an unprecedented effect on the traditional markets of the Ramnas factory and demand for wrought iron declined sharply. The Tersmeden family realised that if the company (and the dependent community) was to survive it would have to find a new market where its skills could be exploited.

The manufacture of chains proved to be just that opportunity and a new factory was built in 1880. The demand for chains steadily increased and they are still one of the major products in the Bulten-Kanthal group.

After the 1943 Bultsfabrik AB takeover the forge was modernised and a secondary factory built on the forge site to manufacture the stainless steel kitchen products.

During the early 1970s it became increasingly difficult to get people to work on some of the dirty manufacturing processes

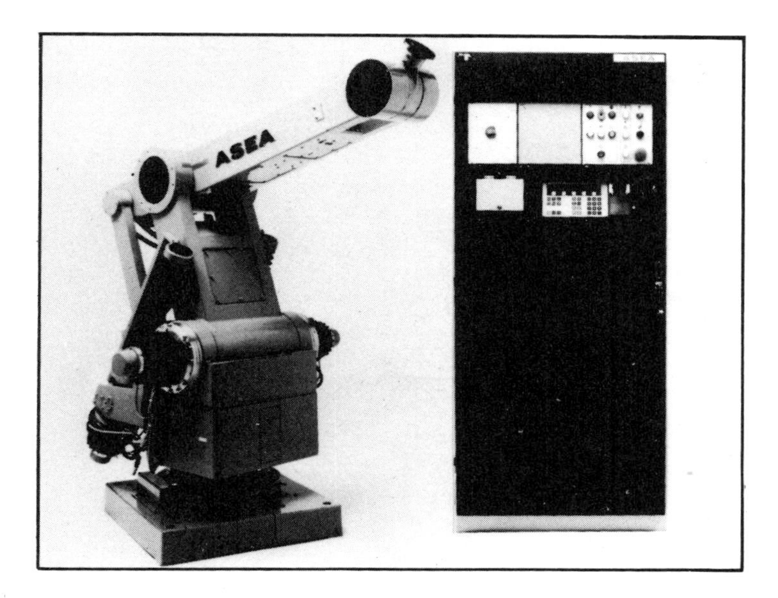


*Fig. 1 The ASEA IRb60 robot and control unit*

associated with the polishing of stainless steel. This, together with the increasing reject rate, prompted the company to consider the use of robots for the polishing tasks. The ASEA IRb60 robot (Figs. 1, 2) was selected for the task.

## The products on which the robot is used

Ramnas Bruk manufactures 14 types of sink tops, 6 types of sink bowls, 3 types of galley sinks, and a variety of other related products including food preparation platters, rinsing bowls, utility sinks, wash tubs, and specialist sinks for disabled people and nurseries. The majority of these products have polished surfaces as well as a variety of texture finishes, particularly on the work surface tops.

The sink tops vary in size from 1000 × 600 mm to 2400 × 600 mm. The sink bowls vary from a single-bowl size of 220 × 340 × 160 mm (deep) to a double-bowl size of 650 × 400 × 160 mm (deep). These products are generally made to standard sizes used in the building trade. However, the units can be customised to suit individual customer's requirements for bowl combinations and outlet holes (Fig. 3).

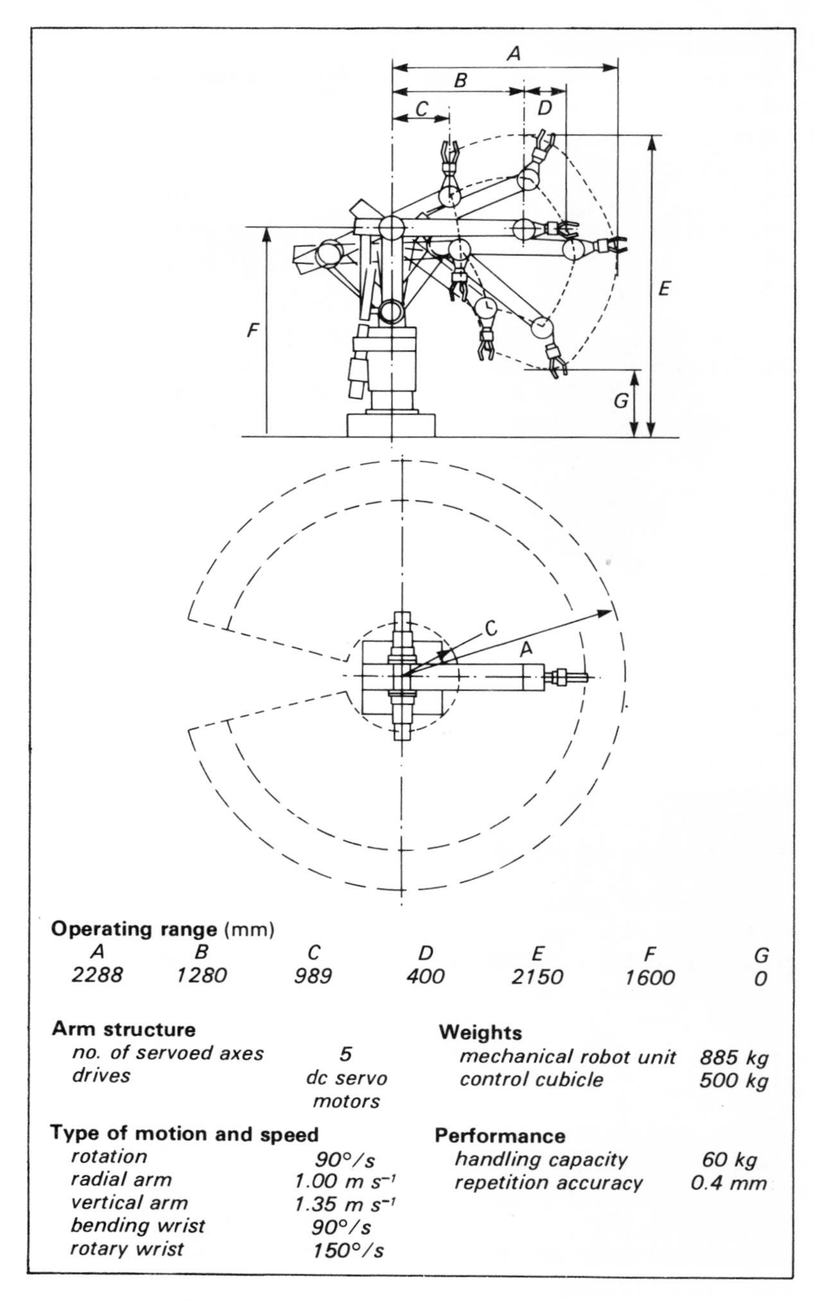


**Operating range** (mm)

| *A* | *B* | *C* | *D* | *E* | *F* | *G* |
|---|---|---|---|---|---|---|
| *2288* | *1280* | *989* | *400* | *2150* | *1600* | *0* |

**Arm structure**

| | |
|---|---|
| *no. of servoed axes* | *5* |
| *drives* | *dc servo motors* |

**Type of motion and speed**

| | |
|---|---|
| *rotation* | *90°/s* |
| *radial arm* | *1.00 m $s^{-1}$* |
| *vertical arm* | *1.35 m $s^{-1}$* |
| *bending wrist* | *90°/s* |
| *rotary wrist* | *150°/s* |

**Weights**

| | |
|---|---|
| *mechanical robot unit* | *885 kg* |
| *control cubicle* | *500 kg* |

**Performance**

| | |
|---|---|
| *handling capacity* | *60 kg* |
| *repetition accuracy* | *0.4 mm* |

*Fig. 2 Technical specifications of the ASEA IRb60 robot*

The manufacturing process for the complete sink tops begins with a stainless steel sheet that is delivered to the factory cut to the right size. The first two processes are to press the sheet to form the tray and to blank out the holes for the bowls that will be fixed later. These processes are carried out on a 800 tonne press. Then the sink-top edges are bent over to stiffen the whole structure. The final process before having the bowls fixed is to weld the corners of the sink to give it added strength and improve its finish.

Sink bowls are manufactured in parallel with the sink tops. Shallow bowls are drawn in a single operation whereas deep bowls are drawn in a double process involving a heating phase between draws. Once drawn, the bowls are trimmed to size in a press. The simple bowls are then passed to the automatic polishing booth where their internal surfaces are polished. The larger more complex bowls are passed to the robot polishing booth. (The automatic booth is a dedicated polishing head that follows a fixed pattern; the robot polishing booth can be changed and programmed to suit the complex contours of other sinks.) If a double sink is required (as in Fig. 3) then the two separate sinks are TIG welded together. The excess metal from this process is then ground away and the joint polished.

*Fig. 3 Double-bowl combination of sink tops*

When the sink tops and the bowls have been completed they are spot-welded together to form the final assembly. Then the structure is passed to the next process where the bowls are seam-welded to the top. The excess metal from the overlap between the components is linished away and the final smoothing and blending of curves is done by hand. The whole sink top is then passed through a polishing station and inspected for any minor blemishes. Finally the sink top is passed to the spray booth where it is given a coat of protective plastic film before being packed and put into store.

## Robot installation and workcycle

The general layout of the Ramnas Bruk robot-polishing installation is shown in Fig. 4. The whole installation occupies about 6 $m^2$. All workstations in the cell are interfaced so that the workcycle is synchronised for each individual operation. The input and output conveyors are also synchronised to the robot's motions and only operate when there is a space in the pick-up position or when a finished component is placed on the output conveyor.

Loading and unloading is done manually; the operator having time to perform any remedial polishing that is required. The input and output conveyors can hold about 15 basins each. The time taken to process a full conveyor is about half an hour and this is long enough for the operator to leave the cell unattended for, say, a lunch or coffee break.

The three polishing machines are used for polishing the sides, the bottom and the drainage slot in the bowl, respectively. Before polishing, the bowl is taken to the first workstation where the polishing wax is applied by a pneumatic gun to the inside of the bowl on all surfaces to be polished. The time taken for this process is determined as a part of the overall robot program for any particular bowl.

The polishing machines have fibre and sisal brushes which need to be replaced after every 80–100 bowls polished. The brush wear associated with the process is accommodated by the robot through a 'soft servo' device. This effectively maintains the pressure between the brush and the bowl at a constant level.

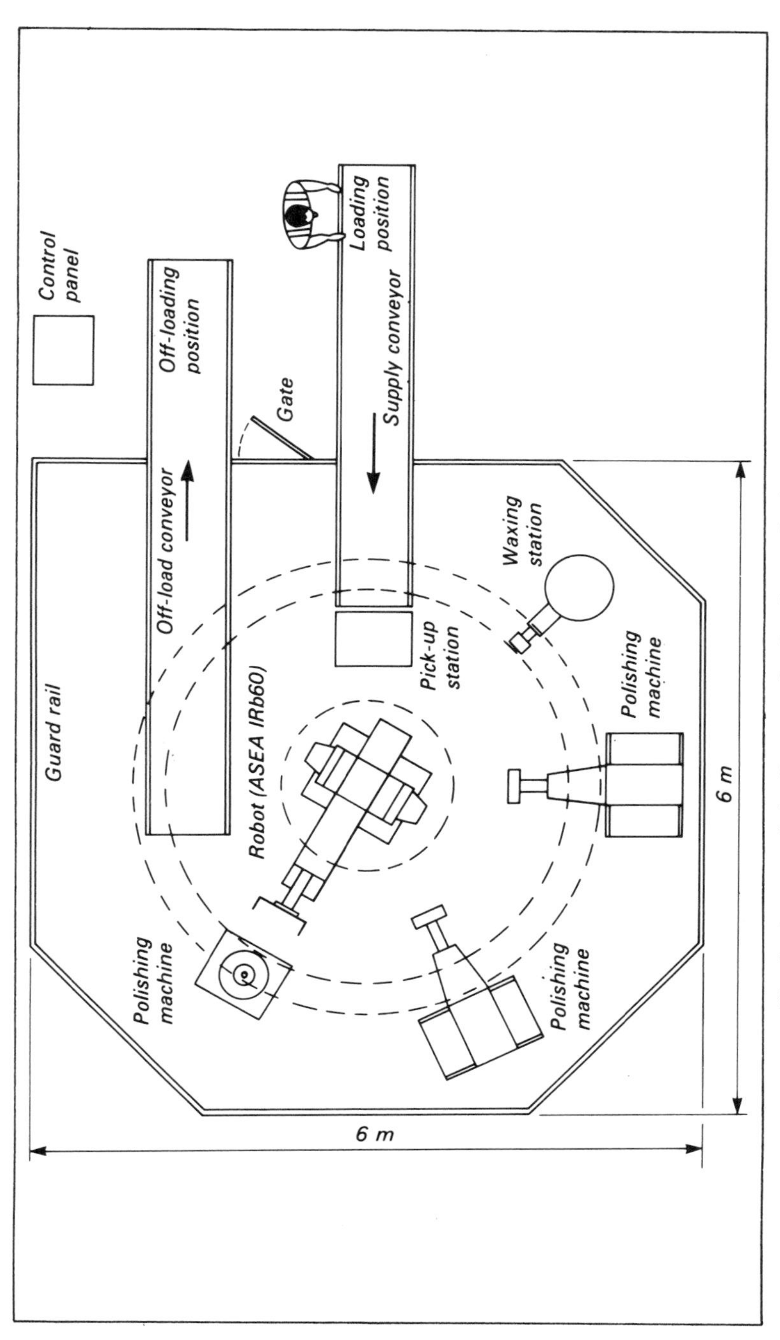


*Fig. 4 Layout of the Ramnas Bruk robot-polishing installation*

The whole cell is surrounded by a safety fence with only one entrance which is interlocked with the robot controller. Should anyone enter the cell, the whole system shuts down.

In total, ten types of bowls are processed through the robot cell. These vary in size from 340 × 220 × 160 mm to 450 × 500 × 340 mm. The batch size varies from 200 to 2000. Depending on size, the robot cell can process between 100 and 260 bowls in an 8 hour shift. This gives a cycle time of between 1.9 and 5 min.

In the normal course of operation only 4% of the bowls require any remedial action from the operator. The cycle time closely approximates to that which is achieved by manual polishing. However, the output from the robot is about 50% higher than that from the manual system due to the robot's continuous output.

**Workcycle description.** The workcycle of the robot cell is more or less the same for all bowls and may be summarised as follows:

(i) The operator picks up the unpolished bowl from the stillage and places it on the input conveyor with the faces to be polished downwards. The conveyor then transports the bowl to the robot loading position.

(ii) Assuming that this position is free (i.e. the robot is processing another bowl) then the sink drops on to the positioning table where a simple pneumatic mechanism centres and squares the bowl ready for the robot to pick it up.

(iii) The robot moves to the positioning table and picks up the sink in specially designed grippers (Fig. 5).

(iv) The robot transports the bowl to the waxing station where the inner surfaces of the sink are sprayed with the polishing wax.

(v) When the waxing has been completed the robot moves the sink to the first polishing station where the inner sides are polished (Fig. 6).

(vi) The robot moves the bowl to the next polishing station where the bottom is polished.

(vii) If required, the sink is then moved to the final polishing station for the drainage slot to be polished.

(viii) Finally the robot moves the bowl to the output conveyor where it is transported to the operator for unloading. The robot then returns to the pick-up station and starts again.

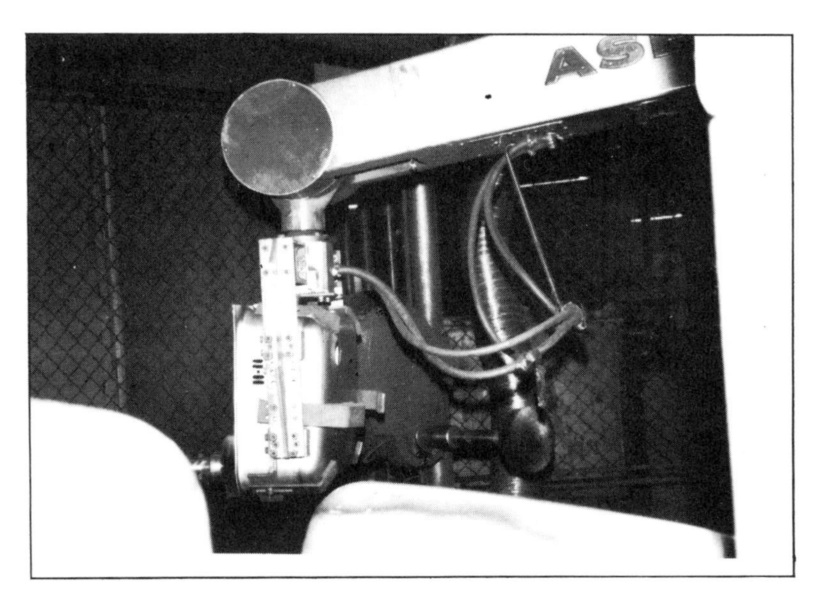

*Fig. 5 Specially designed grippers are used to pick up the sinks*

*Fig. 6 The robot transports the sink bowl to the first polishing station where the inner surfaces of the bowl are polished*

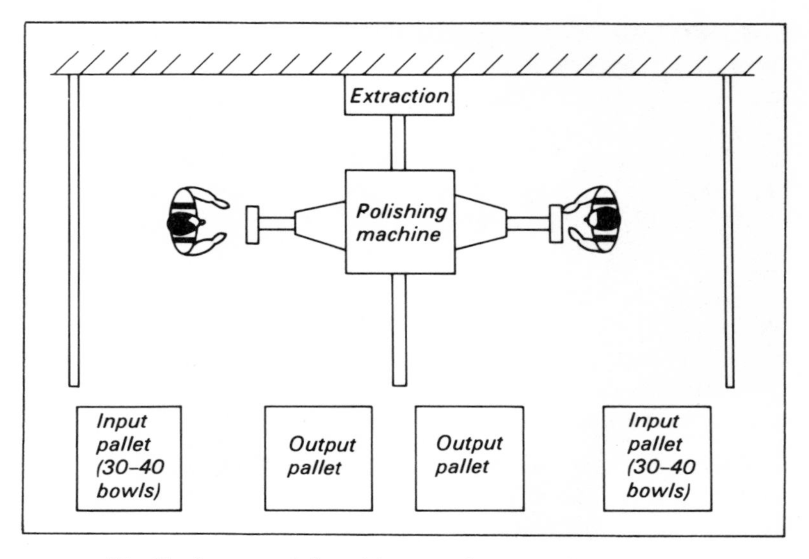


*Fig. 7 Layout of the old manual system for polishing*

## The old manual system

The layout of the manual booths that were previously used to polish the bowls is shown in Fig. 7. The operators worked on either end of a double-headed polishing machine, picking up a bowl from the input stack and then, after applying the wax on the polishing head, manipulating the bowl and polishing it. Wax was applied from time to time throughout the polishing cycle.

The booth did have an extraction unit to remove the dust generated by the process; however, the process was still very dirty. In addition the manipulation of the bowls was a very arduous task; the work was particularly hard on the legs and backs of the operators and absenteeism or sickness was a considerable problem.

## Designing, installing and running the system

Initially the company established a project team to consider how the robot system should be designed and what the performance criteria should be. This team consisted of: the production manager, the project manager, a technical development engineer, an electronic designer, two polishing operators, and an inspector.

Once a potential system had been thought out, Ramnas Bruk contacted three robot suppliers and after some discussions decided that only the ASEA robot could do the job. After consulting ASEA engineers, the Ramnas team and ASEA set up a pilot scheme to consider the detailed design of the system. This began with a test rig set up at ASEA. The robot was programmed and trial polishing operations proved successful.

Initially the idea was to have a robot polishing and a robot picking and placing the bowls. While the trials proved the polishing application they also showed the impracticality of the pick-and-place operation due to the way in which bowls tended to stick together when stacked up.

These trials cost Ramnas Bruk 100 000 SKr (£10 000) and took 6 months to complete. The design of the robot system was finalised and the system subsequently installed in the factory.

A further 4 months elapsed before the system was fully operational with all problems overcome. The majority of the problems were associated with the ancillary equipment and not with the robot. In the five years of operation the robot has only had two serious breakdowns. One was associated with a cable being chaffed due to the continuous movement of the robot, while the other was associated with a control card breakdown.

At present the robot works two eight-hour shifts without interruption. All maintenance is carried out by the company's own technicians.

## Labour relations

As previously mentioned Ramnas Bruk was already experiencing difficulty finding operators for the polishing operation. Once the decision had been made to robotise, the production manager approached the trades union on-site (MIU, Metall Industri Arbefare Forbundel). At first the workforce were suspicious, but they agreed to the introduction on the basis that it would greatly improve their working conditions on the polishing operations.

Since the system was installed there have been no negative reactions to the robot system. This is reflected by the ease of recruitment for people to work with the robot system. In addition the lower number of rejects has greatly improved both productivity and morale.

## Financial considerations

The total cost of the system was 450 000 SKr (£45,000)—about half for the robot and half for the anciliary production equipment. The money was raised by the Bulten-Kanthal holding company.

In justifying the investment Ramnas Bruk could not use conventional payback criteria because of the relatively small batch sizes of 200–2000. However, the company's problems with labour turnover and poor product quality were considered as sufficient justification. With this in mind the payback period was calculated as 4–5 years. In giving the approval Bulten Kanthal waived its usual stipulation of a two-year payback. As it happens the robot polishing cell paid for itself in about three years.

## Important learning points

Since this installation has been working for a number of years the learning points that were originally associated with the system have now been forgotten. However the company does stress that the cost of the original study and the effort expended paid off and it would use the same approach again.

## Future plans

The management of Ramnas Bruk has been considering further investment for some time. Some machine loading and unloading operations have been isolated, and there is also the possibility of using robots in paint-spraying operations.

Many of these plans will have to wait for the general demand of the company's products to improve, but with this in mind the company is still committed to further automation.

**Acknowledgement**

The author gratefully acknowledges the help of Mr. S. Svensson of Ramnas Bruk, Mr. B. Borg of ASEA (Sweden), and Mr. C. Cooper of ASEA (UK), in compiling this case study.

# *Case Study 5*
# Jonas Øglaend AS
## (Sandnes, Norway)

JONAS ØGLAEND AS is part of the Øglaend group of companies located at Sandnes, 10 km to the south of Stavanger in south-west Norway. Jonas Øglaend is the metal division of the group; its main products being bicycles, cable trays and cable ladders. The metal division is divided into the machining factory and the assembly factory. This case study is based on the machining factory; the application being the production of bicycle rims.

The Øglaend group of companies operate in the textile, light engineering, agriculture and real estate markets. In many of these markets they have considerable vertical integration and in the textile market, for instance, they own a number of retail outlets. As a whole the group turned over about 1000 million Norwegian Kroner (£100 million) in 1981 and employed 2300 people. In the metals division the manufacturing plant occupies 7650 $m^2$ and employs about 1000 people. Its turnover in 1981 was 364 million NKr of which the majority was derived from the sale of bicycles.

Those who work at the Øglaend factory tend to live locally and many families have worked in the Øglaend factory for two or three generations. In recent years there has been some immigrant labour, in particular from Vietnam, India and Pakistan, but these only form a small minority of the workforce.

Over 37% of the bicycles produced by Jonas Øglaend and marketed under the brand name DBS were exported in 1981. The company primarily exports to Scandinavian markets, although Finland imports frames rather than complete bicycles. Its share of the Norwegian market, while still dominant at about 75%, has fallen in recent years. This is largely due to an increase in demand that the company has been unable to supply. This has naturally resulted in the import of foreign cycles.

Within the Scandinavian markets Øglaend's main competitor is Monark, a Swedish based company owned by the pop group Abba.

## Company history

Jonas Øglaend first began in 1868 by importing a bicycle called "The World" manufactured by Arnold Schwinn of Chicago. The early years of the business were devoted to the repair and maintenance of bicycles. This changed little until 1906 when Jonas Øglaend began to manufacture bicycle components. Gradually this side of the business began to grow. The range and scope of the spares produced eventually extending to the production of his own cycle.

From about 1920 Jonas Øglaend diversified his business interests. In much the same way as he built up his business in bicycles he bought into small concerns and extended them into the manufacturing areas; the company slowly evolving to its present size and importance.

About 10 years ago the metal division of the company decided to investigate methods of improving manufacturing efficiency. Other Norwegian companies in the locale also had similar thoughts and they all got together to pool resources and to develop some automatic devices. They created an organisation called TESA and between them eventually produced a robot.

Øglaend was not happy with the rate of progress of these early developments and so about six years ago left the organisation. Within a short period of time the company had developed its own robot. This robot (Minimater) and its subsequent derivatives are now widely applied throughout Øglaend's own factories and of course they are manufactured under license abroad.

UK distributors of the Øglaend robots are Fairey Automation of Swindon. The Minimater (Figs. 1, 2) is a general purpose handling robot designed to cover applications

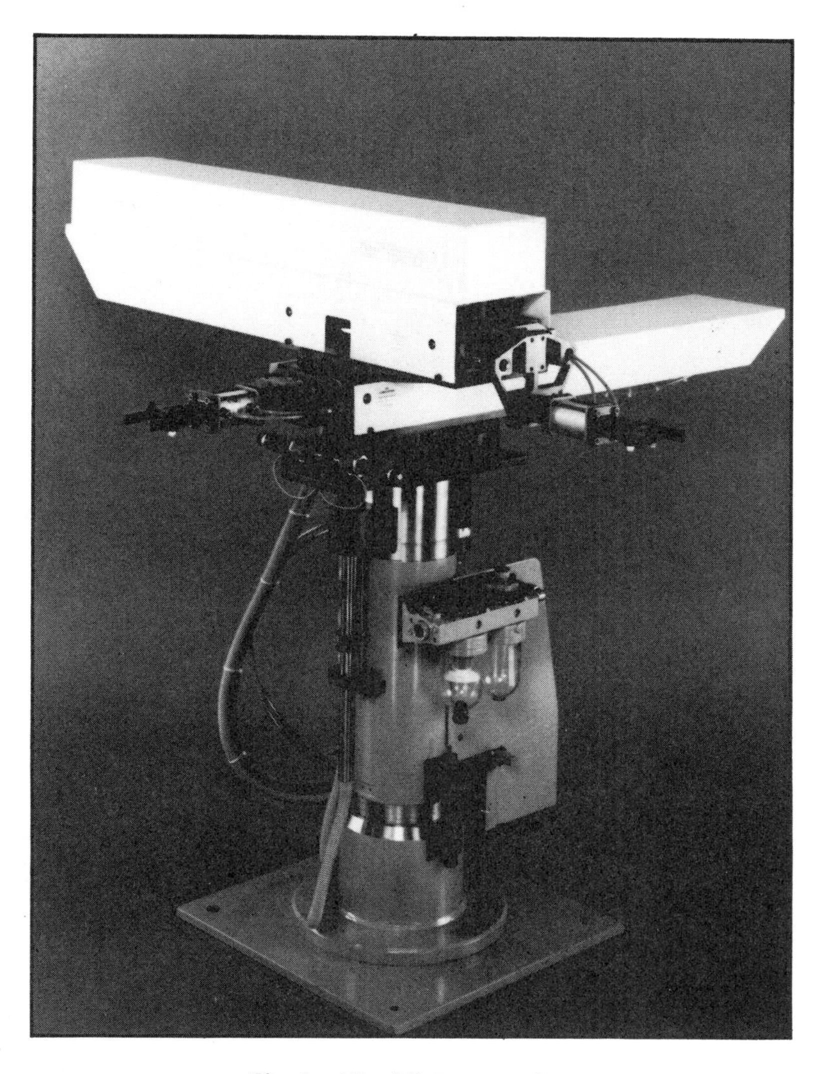

*Fig. 1 The Minimater robot*

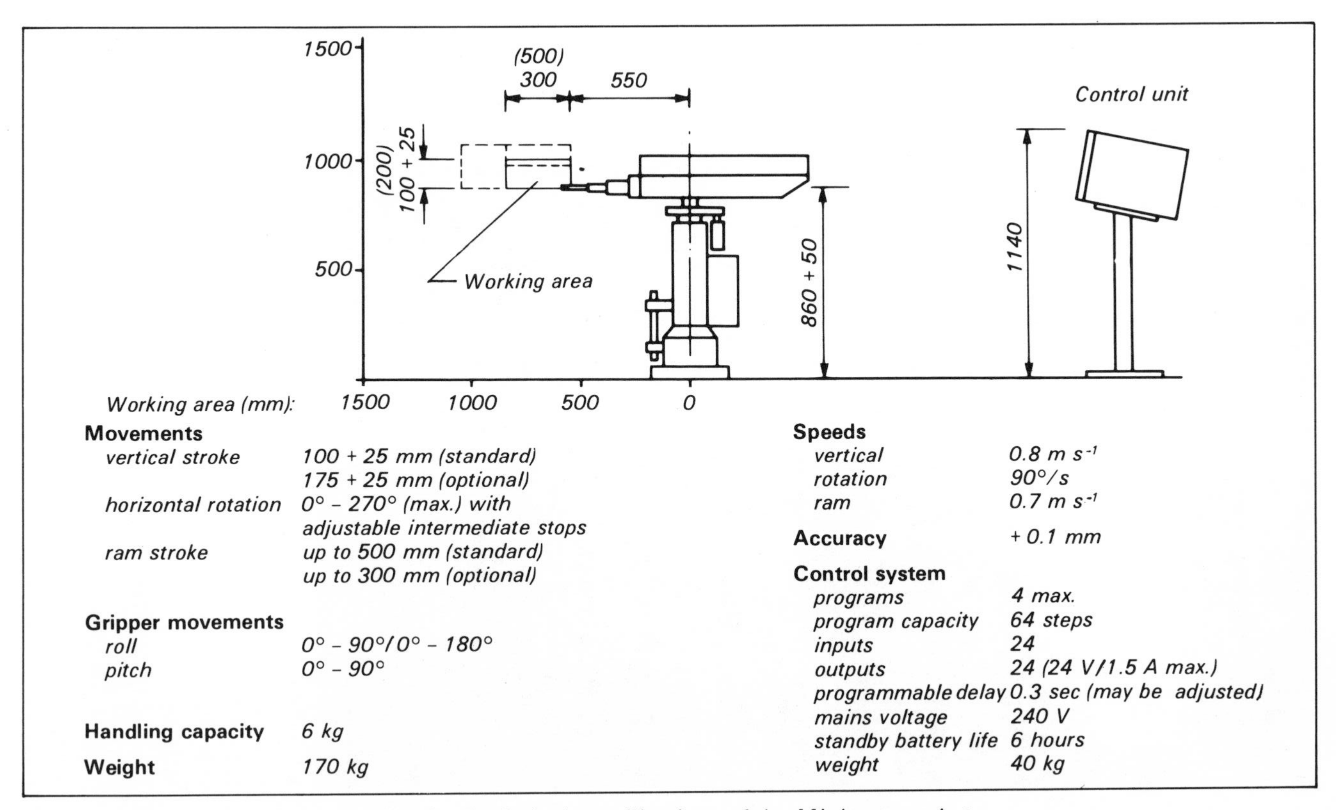


Fig. 2 Technical specifications of the Minimater robot

which require light parts or components (up to 6 kg) to be manipulated. The robot is pneumatically operated and uses adjustable stops to provide the programmable operating positions. Standard grippers are available which allow roll (0°–90° or 0°–180°) and pitch (0°–90°) movements. Loads may be gripped mechanically, magnetically or by suction pads.

## The products on which the robot is used

As previously mentioned robots are being used extensively throughout the whole of the production of the DBS bicycle. The functions the robots perform range from simple pick-and-place tasks to arc-welding. Not all of these are Øglaend robots, for instance Trallfa robots are used for paint spraying.

Since so many types of bicycle are made it would be impossible to describe all production methods. Therefore it is simpler to describe one and to assert that there is little basic difference in the manufacture of others.

Bicycles are made in a series of subassemblies, conveniently grouped as:

(a) the framework,
(b) the wheels,
(c) the foot cranks, bearings and gears,
(d) the ancillary equipment including tyres, chain, dynamo, seat, electrics, cables and brakes.

Nearly all of the items in part (d) are bought-in from other companies both home and abroad. The framework and the wheels are made in two stages. First, the manufacture of the individual components such as framework tubes or cranks, and secondly the assembly of these components. As already mentioned the two stages are presently performed at different sites. In the manufacturing factory the framework tubes are cut to length, have their ends shaped, and have any holes drilled in them that are required. Tubes that are used for items such as the handlebars are cut to length, bent to shape and subsequently chromed before being despatched to the assembly factory.

The wheels are made from an aluminium alloy extrusion supplied in 5 m lengths. This extrusion is rolled into a spiral of the required wheel diameter, and then cut. The individual coils are fed into the robot cell for butt-welding and the subsequent machining processes. The rims are then polished and anodised before being passed on to the assembly factory.

Foot cranks (bought-in as forgings) are de-flashed and subsequently machined to take the pedals and bearings. The cranks are then chromed and despatched to the assembly factory.

The primary gear is punched from sheet metal, then machined, hardened and chromed.

At the assembly factory the frame components are assembled and either welded or braised into position. The

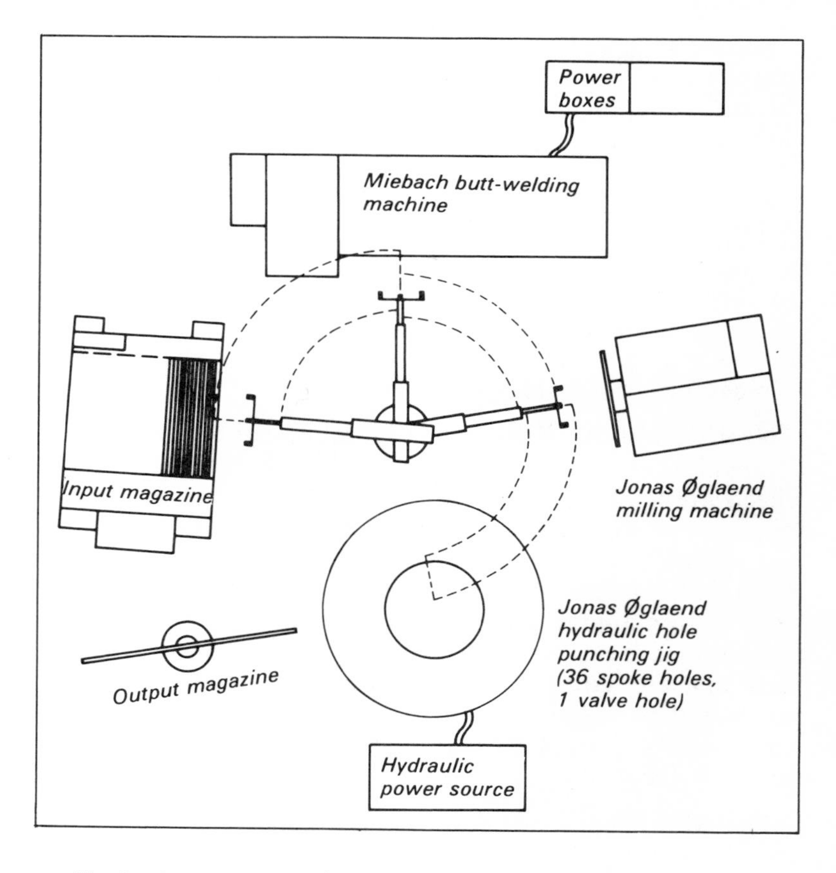


*Fig. 3 Layout of the Øglaend robot rim-manufacturing station*

frame is then cleaned and painted in batches of 1–300 (normally each batch will be a single colour). The wheels have the spokes and hubs fitted and tensioned; then the inner tubes and tyres are fitted. Finally, all components are assembled into the complete bicycle, packed and despatched. (Note, this assembly factory is serviced by an automatic storage system.)

## Robot installation and workcycle

The general layout of the robot cell for bicycle rim manufacturing is shown in Fig. 3. Three Minimater robots are mounted on a single common pivot; the relationship of these arms is determined by the surrounding machinery. This robot services four workstations.

At the first workstation the first robot arm picks up an unwelded rim from the input magazine (Fig. 4) and takes it to the butt-welding machine (Miebach 50 kWA) where it is clamped, and the robot arm retracts. The butt weld is then formed.

Once the butt weld has been completed the second robot arm picks up the rim and takes it to the milling machine (specially designed by Jonas Øglaend). Here the rim is clamped, the

*Fig. 4 The robot arm picks up the rim from the magazine*

*Fig. 5 The bicycle rim is clamped to the milling machine*

*Fig. 6 The milling machine mills the weld both inside and outside the rim simultaneously*

*Fig. 7 Circular jig hydraulically punches 36 spoke-holes and an air-valve hole into the rim*

robot arm retracts, and the machine mills the weld flush both inside and outside the rim simultaneously (Figs. 5, 6).

Finally the third robot arm picks up the rim from the milling machine and places it on the fourth workstation. Here the rim is placed into a circular jig and the 36 spoke-holes plus an air-valve hole are punched simultaneously into the rim. The punches are hydraulically operated (also designed by Jonas Øglaend, Fig. 7). The rim is then ejected from the workstation by a simple lever mechanism that flicks the rim on to the output stack.

The speed of this process can be adjusted from a minimum of three minutes to a maximum of four. The operation cycle time is about 50–54 seconds.

## The old manual system

The manual system of producing the bicycle rims was rather like the present robot system except that it was laid out in a linear manner. One operator was required per workstation and between each workstation there was a buffer stock. The difference in the work processes only really relates to the final hole punching operation. In the manual system the holes were punched two at a time and of course this took considerably longer than the present system.

## Designing, installing and running the system

The robot cell was designed by Øglaend's own engineers and was among the first to exploit its own robots. The actual layout of the cell was designed by a small team who included the foreman of the department and two operators who were later to operate the system. (This approach has also been used in subsequent robot applications within the organisation.)

The original design of this workstation evolved from a similar workstation that had three fixed automation-type work arms working under a single cover. This system was very difficult to maintain and adjust, thus it was natural to apply the new technology in an almost tailor-made situation.

The new system was installed close to the original system and 'run-up' over about four months. The major difficulties encountered during the running-in period were associated with the machine alignments and with the clamping of the rims at the various workstations. In particular the butt-welding station proved difficult to service as the metal vapour released during the butt welding operation tended to be deposited in the clamping mechanism. Eventually this problem was overcome by the automatic application of a lubricant before the rim is placed and clamped in the machine.

By and large the robot installation is relatively trouble-free in operation. The whole system is regularly maintained and at every works holiday shut-down it is overhauled, adjusted and recommissioned.

## Labour relations

The workpeople of Øglaend are unionised and belong to the JM (Jern og Metal). When the robot systems were being designed the union was aware of what was going on and it came as no surprise when the company began to introduce robot systems.

Initially the individual operators who had not been involved in the design of the robot system were suspicious. As time has passed, so they have accepted the use of robots in general and there is no trouble. Perhaps the fact that Øglaend is also manufacturing the robots also encourages fairly positive attitudes. More generally, in normal operations the only antagonism that is ever generated is if the robot breaks down, as it threatens the group bonus for the operators concerned. In these circumstances the union *insists* that the robot is repaired immediately.

## Financial considerations

The system originally cost about 500 000 NKr (£50 000) and the machine was justified by an increase in productivity. This was achieved through the use of one operator instead of three and a reduction of work-in-progress. Details other than these are not available.

## Important learning points

Øglaend asserts that although this system is simple in concept it is quite difficult to convert this concept into reality. The company's general experience with this and other systems tends to show that the relationship between the complexity of the system and the development time is not linear. This is largely because some of the problems that are encountered when the system is run up are difficult to anticipate. Because of this past experience Øglaend now allows longer development times for its systems.

## Future plans

The company has already introduced robots into many areas of its factories and this process will continue. At present a number of new robot applications are nearing completion. Plans for the future are to continue this process where it is sensible and economically viable to do so.

In addition to the internal development of robot uses and devices, the company also wishes to extend the number of countries where its robots may be made under license.

**Acknowledgement**
The author thanks Mr. Hana of Jonas Øglaend and Mr. R. Shields of Fairey Automation for their assistance in compiling this case study.

# Case Studies— Summary and Analysis

THE PREVIOUS five illustrations of robot applications provide a good opportunity to compare how differing manufacturing traditions cope with the introduction of robots. The aim has not been to compare similar robot applications but to examine applications in different countries to see where common factors in the use of robots exists.

With this in mind the following questions serve to provide a focus for an analysis and to provide a vehicle for the introduction of other information which could not be mentioned in the context of the studies:

- How did the companies arrive at the point of robotisation and what affected their choice of robot?
- How did the companies design their robot systems and how long did they take to become productive?
- What kind of problems were encountered in the running-in of the systems?
- What have the trade unions and workers attitudes been towards the introduction of robots?
- How have the systems been economically justified?
- How has the experience of robots affected management attitudes towards automation?

In examining these issues the case studies become a useful adjunct to the largely technical first part of the book.

## How did the companies arrive at the point of robotisation and what affected their choice of robot?

An examination of the histories of the different companies involved in the case studies shows no distinct common pattern of evolution leading to robotisation. The majority of the companies have been through a phase of single ownership to emerge as a part of a larger industrial combine—but this only seems to have been of any significance in the raising of capital (this will be explained later).

If there is any common feature to be found then it must relate to the positive attitudes of the managers and company executives who made the decision to invest in robots. Underlying these decisions has been a variety of motives ranging from market pressures to the need to replace people in boring, arduous or dirty working environments.

The case studies were chosen quite deliberately so that a French company was using a French robot, an Italian company using an Italian robot, and so on. Again, the circumstances that led each company to choose their particular robot have nearly always resulted from the same kind of evaluation process. The final choice of robot seems in part to have been influenced by the ability of the robot to do the job and in part by the reassurance of dealing with a robot company from the same country.

However, this is not in itself of great significance as other companies that could have been chosen for the illustrations have used foreign robots. Thus it is difficult to be definitive about what influences managers to buy a particular robot. The only comment that can be made is that if two robots—one from the same country and one from abroad—are capable of doing the same job to the same standards and with the same levels of support, most managers and engineers prefer the 'local' robot.

## How did the companies design their robot systems and how long did they take to become fully productive?

The majority of the companies studied managed to complete the outline design of their own robot systems. The subsequent detailed design stages were then usually done by the robot suppliers or robot consultants.

The team of people involved in the outline design of the system has varied from a production engineer and a manager, to almost a vertical cross-section of people in a company from manager down to operator. The managers who included their operators asserted that they played a positive role. The choice of whether or not to include people from the shopfloor was largely a function of the prevailing management style of a company. Clearly companies are best to draw on their own resources whenever possible.

None of the systems that were examined became fully productive immediately on installation. Each system has had a running-in phase. The more complex the system the longer the running-in phase. The usual time taken for a system to settle appears to be between two and six months.

## What kind of problems were encountered during the running-in phase of the robot systems?

It is very significant that few complaints about robot reliability were voiced. The majority of problems encountered were associated with the ancillary equipment and the jigs and fixtures. In more than one instance production engineers have said that they could not have anticipated all the problems that they encountered.

There are two probable explanations for these comments. The first is related to the lack of experience that most companies have in the application of robots. (There is no doubt that a few engineers are familiar with the versatility and the limitations of robots.) The second reason lies in the fact that much production equipment is designed with human workcycles in mind. Robots tend to stress this equipment more highly and as a consequence it breaks down more frequently. In other applications it is often found that jigs and fixtures have to be redesigned to cope with the need for robot access.

While many of these problems have been trivial they have taken time to overcome. To avoid them the only solution would be to spend more time in the analysis and design phase, and with more expert help.

## What have the trade unions' and workers' attitudes been to the introduction of robots?

In every case study the trade unions and the workforce have been suspicious of managements' motives in the introduction of new technology. However, in every case—subject to suitable safeguards regarding jobs—they have not opposed the introduction of robots. In addition, they have often quite freely adapted existing manual work methods with the minimum of bother.

These findings are quite remarkable, for this is an area of great concern for managers. It would seem that no matter what the trades union traditions in any country, the overall response has been positive. This suggests that workers and trade unions take a fairly pragmatic view of technological developments. How far these attitudes would prevail in the face of extensive automation cannot be judged from these studies. However, it is worth noting that the majority of the managers did involve their workpeople in consultation and education at an early stage in the planning of the robot systems.

## How have the systems been economically justified?

Some quite different reasons have been used to financially justify the use of robots. It would appear that robots are most easily justified when there is a large volume of homogeneous products being made. In the two cases studies where large production runs were involved, the robots have resulted in large labour and work-in-progress savings. These economies, together with higher quality, have ensured that the systems have paid-back well within the prescribed time.

In other cases, where robots have been justified on health or environmental grounds, the payback period has been longer. This is particularly true as the batch size becomes smaller and the variety of batches increases. However, even in these cases the manufacturing efficiency has increased to levels that were simply not attainable with manual labour. As the batch size decreases then the justification of the robot *per se* becomes less important and more attention has to be paid to the surrounding equipment.

The final category of financial justification is centred on the need to achieve economic growth. In these circumstances the concept of payback is less important than the long-term return on capital employed. The issues in this approach are more cloudy as more strategic issues tend to be raised. These are notoriously difficult to comment on as they are very much a question of management beliefs.

What is interesting is the number of companies who have used some internal source of funding rather than going to the banks. There is no explanation for this and one must only assume, rightly or wrongly, that the traditional banking system may be too conservative in its outlook!

## How has the experience of robots affected management attitudes toward automation?

The companies and management in the case studies all risked the introduction of robot systems. In some cases the failure of the system could have been catastrophic. The success of the systems is undoubtedly a tribute to them and their workforce.

In all cases the first experiences with robots encouraged the management to either be considering the further use of robots or automation, or to have actually gone ahead with further investment programmes.

These questions and the results of the case studies clearly illustrate the international nature of robot technology. Their use transcends boundaries and cultures and their success seems only to be dependent upon the vision of managers and upon the skill of production engineers and project managers.

# *Appendix A*
# Interface descriptions

THE FOLLOWING tables are illustrations of the relationship between the robot and the equipment with which it works (see also Chapter Three). Each begins with one of the basic classifications of robot usage. Then the most common batch size processed by the robot in this configuration is described. Again the classification is related to its normal manning levels (so 1:1 would mean one man to one robot; 1:3 one man to three robots). Under the general classification the most common applications are identified. In the tables below this information the main system component, other than the robot, is identified (e.g. manipulator—grinding and arc welding). The accompanying columns show what will be functionally required between the robot and the equipment (columns 2 and 4), and column 3 shows the direction of the interface.

So, in the first table it is found that in the grinding or arc welding applications of the robot, it is usual to have a work manipulator. When the robot program begins there will have to be a 'cycle initiate' signal from the manipulator to the robot to tell it that it has finished moving and is in a stable state. This signal is either a direct digital signal (where the manipulator has its own controller) or an electrical switch. On receipt of the signal the robot will begin its cycle.

If either the robot controller or the manipulator controller detect an error then a 'cycle abort' signal will be generated. This can come from either the robot or the manipulator and can pass in either direction. With this notification it is possible to see how the different systems will be interfaced and the type of interface that will be required.

*Robot classification*—robot as a tool user and working in a stand-alone application.

*Batch size*—very small, small and medium

*Manning levels*—1:1 and 1:3

*Applications*—flame cutting, grinding, arc welding, marking

| System components | Robot interfaces with equipment | Direction and type of interface | Equipment interfaces with robot |
|---|---|---|---|
| Process (e.g. equipment controlling process | Electrical switch button | ← Cycle initiate (manual) → | Electrical switch button |
| | Electrical switch | ← Emergency stop → | Electrical switch |
| | Software trigger | ← Digital signal → | Process malfunction |
| | Robot malfunction | Digital/electrical switch | Process shutdown |
| Manipulator (grinding and arc welding) | Cycle initiate | Digital/electrical switch → | Start cycle |
| | Cycle abort | ← Digital/electrical switch → | Cycle abort |
| | Electrical switch | ← Emergency stop → | Electrical switch |
| Tool (grinding and marking) | Software trigger | ← Digital signal → | Tool wear via sensor or measurement |

Note, ⟷ indicates the direction of the interface

*Robot classification*—robot as a work handler and working as part of a production line

*Batch size*—large and very large

*Manning levels*—1 : many

*Applications*—loading, palletising, stacking, packing, sorting

| System components | Robot interfaces with equipment | Direction and type of interface | Equipment interfaces with robot |
|---|---|---|---|
| Conveyor(s) | Speed synchronisation | Digital analogue optical mechanical □ | Speed sensor |
| | Electrical switch | ← Emergency stop → | Electrical switch |
| | Electrical switch/ digital signal | ← Start stop → | Electrical switch/ digital signal |
| | Analogue/digital signal | ← Speed adjust → | Analogue/digital signal |
| Gripper mechanisms | Cycle test and initiate software | Digital analogue optical | Touch/position sensors |

*Note,* □ *indicates that the interface can be of different types (e.g. digital, electrical or optical)*

*Robot classification*—robot as a tool user and working as a part of a production line

*Batch size*—large and very large

*Manning levels*—1 : many

*Applications*—paint spraying, finishing, spot welding, sealing, inspection

| System components | Robot interfaces with equipment | Direction and type of interface | Equipment interfaces with robot |
|---|---|---|---|
| *Conveyor* | *Speed synchronisation* | *Digital analogue optical mechanical* □ | *Speed sensor* |
| | *Electrical switch* | ← *Emergency stop* → | *Electrical switch* |
| | *Electrical switch/ digital signal* | ← *Start stop* → | *Electrical switch/ digital signal* |
| | *Analogue/digital signal* | ← *Speed adjust* → | *Analogue/digital signal* |
| *Process or tool status* | *Tool/process software trigger* | *Digital analogue electrical* □ | *Out of tolerance* |
| | *Tool replacement software cycle* | *Digital analogue* □ | *Tool replacement sensor* |
| | *Electrical switch* | ← *Emergency stop* → | *Electrical switch* |
| | *Electrical switch/ digital signal* | ← *Start stop* → | *Electrical switch/ digital signal* |

*Robot classification*—robot as a work handler and working in a stand-alone application

*Batch size*—very small, small and medium

*Manning levels*—1 : 1 to 1 : 3

*Applications*—forging, fettling, investment casting

| System components | Robot interfaces with equipment | Direction and type of interface | Equipment interfaces with robot |
|---|---|---|---|
| Process equipment (forge furnace) | Electrical switch/ button | ← Cycle initiate (manual) → | Electrical switch/ button |
| | Electrical switch | ← Emergency stop → | Electrical switch |
| | Software trigger | ← Digital signal | Process malfunction |
| | Robot malfunction | Digital/electrical switch → | Process shutdown |
| Gripper mechanisms | Cycle test and software switches | Digital analogue □ optical | Touch/ position sensors |
| Tool | Software trigger | Digital signal → | Tool wear via sensor or measurement |
| Manipulator | Cycle initiate | ← Digital/electrical switch → | Start cycle |
| | Cycle abort | ← Digital/electrical switch → | Cycle abort |
| | Electrical switch | ← Emergency stop → | Electrical switch |

# *Appendix B*
# Model New Technology Agreement

THE Model New Technology Agreement (as discussed in Chapter Five) is included as a sample which reflects many of the commonly found trades union responses to particular issues. Different trade unions have different specific guidelines depending on the nature of their members' occupations. This model, therefore, must not be taken as representing a specific response.

**General**

1. This is an agreement between the Management and the Union covering technological change, insofar as it affects or might affect the Union membership, either individually or collectively.

2. The term 'technological change' in this agreement shall be taken to include all changes or proposed changes in equipment, materials, processes or products; and all changes or proposed changes in working practices or conditions associated therewith.

3. The Management and the Union agree that technological change will only be successfully achieved if proper regard is paid at each stage to the principle of job security and control.

4. This agreement should be read as a whole.

**Procedural**

5. Where technological change results in a disagreement between the Management and the Union, working practices shall revert to what they were prior to the disagreement, and the change shall not be made until it is agreed through the negotiating procedure.

6. The Union reserves the right to re-open negotiations on existing agreements where their application is altered through subsequent technical change.

7. This agreement shall be subject to annual review.

**Job security**

8. There shall be no reduction in overall employment levels as a result of technological change.

9. The Management undertakes to bring forward a programme for future employment, including targets for growth in employment levels.

10. Any reductions in standard labour input as a result of technological change shall wherever possible be managed through agreed reductions in working time without loss of earnings. Negotiations shall give full consideration to the following methods of reducing working time:

(a) reductions in normal working hours,
(b) elimination of overtime,
(c) increased holidays,
(d) voluntary early retirement with full pension rights,
(e) sabbatical leave,
(f) time off for training or retraining.

11. No worker shall be downgraded or suffer a financial loss as a result of technological change.

**Job control**

12. The Management and the Union agree to develop machinery and procedure appropriate to their joint control of technological change.

13. A committee shall be established, composed of Management and Union representatives, to consider all matters arising in connection with technological change.

14. The committee shall in particular concern itself with the following aspects of technological change:

(a) investment and research and development programmes,
(b) equipment and process design, location and installation,
(c) product design,

(d) all manpower issues arising, including employment and manning levels, training and retraining,
(e) monitoring of all aspects.

These matters shall be considered by the committee and agreement reached on all substantive issues, in all cases prior to the ordering of new equipment.

15. The Union reserves the right to take through the established disputes procedure any matters coming before the committee. The exercise of this right shall have the effect of activating the provisions in clause 5 above, and of suspending discussion on the items in discussion by the committee.

16. The Management shall bring all proposals for technological change to the attention of the committee at the earliest possible stage, and shall disclose sufficient information to enable Union representatives to understand the fundamental features of the changes proposed, and the importance of these changes to the Union members.

17. The Management shall disclose to the committee all information requested by the Union in connection with technological change and with the fulfilment by the committee of the particular functions described in clause 14 above.

18. The Management undertakes to enter into a disclosure of information agreement with the Union.

19. The Management shall appoint a secretary to the committee who shall handle all the necessary administrative and clerical work. Agreed minutes shall be kept of all committee meetings and shall be made available to all committee members.

20. The committee shall meet at the request of either the Management or the Unions, and in any case shall not meet less than once a quarter.

21. Committee meetings shall be held whenever possible during working hours. Union representatives shall be given time off at average earnings to attend.

22. The Management shall extend to the Union side members of the committee the facility to attend Union approved training courses related to technological change. Time off shall be paid at average earnings. The Management agrees to enter

into an agreement with the Union providing time off for training.

23. Union representatives on the committee shall be given full facilities for reporting committee proceedings and decisions to their membership; the use of such facilities shall involve no financial loss to any Union member.

24. Full-time Union officials and advisers shall be eligible for co-option by the Union side of the committee.

25. The collection, storage, processing or use of personal data in computer systems shall not take place unless it can be justified as being necessary for efficient management. The Management undertakes to enter into an agreed procedure governing all aspects of the collection, storage, processing and use of personal data.

**Health and safety**

26. Safety representatives shall be involved at the earliest possible stage in the conception and planning of technological change. It is recognised that the Health and Safety at Work Act and the Safety Representatives' Regulations give safety representatives the right to be involved at this earliest stage.

27. It is most effective, and in the interest of Management, to eliminate hazards at the design and planning stage. It is also recognised that the views of the final users must be made known to designers and planners if all foreseeable risks and needs are to be identified. The Management therefore agrees to involve safety representatives in agreeing specifications with architects, designers and planners.

28. The Management undertakes to:

(a) monitor all available literature and other sources of information on hazards and technical developments,
(b) supply this information to safety representatives, and
(c) act on information which safety representatives consider important.

29. The Management undertakes to identify and, so far as is possible eliminate any ill effects of technological change on all those who may be affected. This may involve obvious hazard

removal, but may also require changing the application of technology, cooperating with other employers to create jobs, or prohibiting imports from countries with poorer working conditions that in the UK.

30. The Management undertakes to provide medical surveillance and meet any costs of prevention and treatment that may be necessary to particular hazards.

31. Safety representatives shall be allowed paid day release for union training, and shall be provided with appropriate technical training by the Management so that they can be more effective in joint consultation. This means that courses should take place before detailed consultation over the design of technological change begins.

# *Appendix C*
# Financial appraisal case study

THE EXPLODED view of a 13 Amp plug whose assembly has been chosen by the manufacturer as a suitable task for a robot is shown in Fig. 1. This application has been chosen as a pilot scheme. If it is successful, similar assembly tasks elsewhere in the factory will be converted. Although hard automation has been considered, the management wanted to introduce a new plug design within twelve months and it was felt that retooling would be too expensive. The precise timing of this new plug could not be anticipated so it was decided to use the existing plug as the basis for calculations.

With manual production techniques the plug takes 6 min to assemble; this includes screwing the whole unit together. In a production shift the normal rate of work gives an output of 59 plugs per operator. With the introduction of a bonus system this has been increased to 65 plugs per operator. The factory works 270 days a year on average (including some weekend working to cover production peaks), and two shifts are worked per day. Thus the number of plugs produced per manual assembly station per annum is:

$$65 \times 2 \times 270 = 35\,100$$

The average selling price of a plug is, say, £1.50 giving an annual revenue for each workstation of:

$$35\,100 \times £1.50 = £52\,650$$

Preliminary investigations by the company's production engineers show that a cycle time of 1.5 min should be achieved by the robot. In addition the overall output increases to 300 per

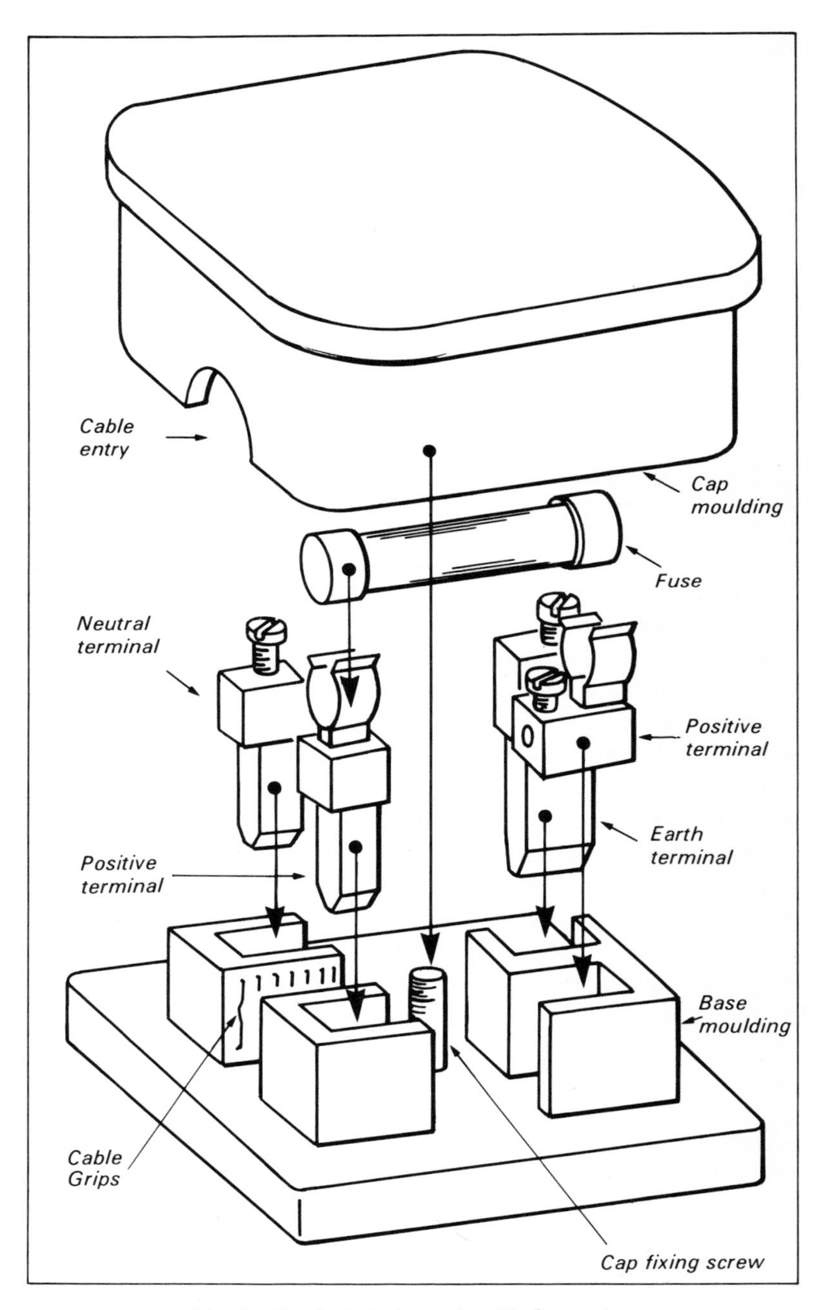


*Fig. 1 Exploded view of a 13 Amp plug*

shift. The following table shows how this figure was derived:

| *Shift hour* | *Manual productivity* | *Manual output* | *Robot output* |
|---|---|---|---|
| 1 | 90% | 9 | 40 |
| 2 | 100% | 10 | 40 |
| 3 | 100% | 10 | 40 |
| 4 | 60% | 6 | 40 |
| 5 | 0 | 0 | 40* |
| 6 | 80% | 8 | 30† |
| 7 | 100% | 10 | 40 |
| 8 | 60% | 6 | 30‡‡ |
| | | 59§ | 300 |

* *It is assumed that the robot will work throughout the lunch break with full magazines and the minimum of manual intervention*
† *Output will drop in this hour due to magazine reloading*
‡‡ *Output will drop in this hour due to the need to fill the magazines for the following shift*
§ *With bonus this figure is raised to 65*

Thus by comparison the robot productivity is 300 ÷ 65 = 4.6 times that of the manual system.

The plug production line employs ten operators and on the basis of the above calculation the management expects to replace nearly all of the existing operators on the line. To be certain that contingency is built into the line for items such as breakdowns and maintenance, three manual workstations are to be retained. Thus on each shift there will be three operators on manual production and one tending the robots.

Overall the savings that are achievable using two robot systems are six operators per shift, or 12 operators over the normal working day. The company reckons that the annual cost of an operator is about two times their salary, i.e. 2 × £5 000 = £10 000 per annum (this includes all relevant overheads). Using this figure the total saving per annum is 12 operators × £10 000 = £120 000. The company requires a payback period of three years or less and return on capital invested greater than 30%.

Against these projected savings there are the following costs:

| *Expenditure* | *Costs (£)* | |
|---|---|---|
| *Manufacturing equipment* | | |
| *basic robot packages (2)* | *70 000* | |
| *grippers (2)* | *7 500* | |
| *fixtures (2)* | *15 000* | |
| *interlocks and interfaces (2 sets)* | *5 000* | |
| *process equipment (2 sets)* | *20 000* | |
| | *117 500* | *117 500* |
| *Systems engineering* | | *10 000* |
| *Structural modifications and equipment* | | *10 000* |
| *Safety* | | |
| *enclosures* | *6 000* | |
| *alarms* | *500* | |
| *operator protection* | *3 000* | |
| *training* | *1 500* | |
| | *11 000* | *11 000* |
| *Personnel (redundancies and training)* | | *20 000* |
| *Project management* | | *5 000* |
| *Feasibility study* | | *5 000* |
| *Maintenance spares* | | *5 000* |
| *Product redesign* | | *25 000* |
| *Learning (say 0.5% lost production)* | | *26 300* |
| | | *£234 800* |

*Note, there are no enhancements required on the basic robot package, no work positioners, and no computer hardware and software*

Negotiations with the Department of Industry indicate that a 25% grant will be available. The company is using a 15% DCF (see Chapter Six) for all current investment appraisals. It is estimated that the life of the robot system is five years and the company write-off rate is 25% per annum. The residual value of the system is estimated at £7000.

Parameters may be summarised as follows:

| | |
|---|---|
| Total systems costs | £234 800 |
| Annual savings generated | £120 000 |
| Life of system | 5 years |
| DCF factor | 15% |
| Capital write-off rate | 25% |
| Residual value | £9286 |
| Government grant (25%) | £58 700 |
| Assume tax | 48% |

*Stage 1*—Cash flow calculations

| | | | Cash value at 48% tax |
|---|---|---|---|
| *Cost of robot system* | | *234 800* | |
| *Less annual write-off* | *58 700* | | |
| *Less government grant* | *58 700* | | |
| | *117 400* | *117 400* | |
| | | *117 400* | |
| *Year 1 write-off allowance (25%)* | | *29 350* | *14 088* |
| | | *88 050* | |
| *Year 2 write-off allowance (25%)* | | *22 012* | *10 566* |
| | | *66 038* | |
| *Year 3 write-off allowance (25%)* | | *16 510* | *7 925* |
| | | *49 528* | |
| *Year 4 write-off allowance (25%)* | | *12 382* | *5 943* |
| | | *37 146* | |
| *Year 5 residual value* | | *9 286* | |
| | *Balancing allowance* | *27 860* | *13 373* |
| | | | *£51 895* |

From the calculations in Stage 3 it can be seen that the payback period of this investment is about 15 months and that the return on capital invested is 58%. This meets both of the company's financial criteria and is therefore acceptable as an investment.

*Stage 2*—Cash flow summary

| Year | 0 | 1 | 2 | 3 | 4 | 5 | 6 |
|---|---|---|---|---|---|---|---|
| Capital cost | -234 800 | | | | | | |
| Residual value | | | | | | | 9 286 |
| Government grant | | 58 700 | | | | | |
| Write-off | | 58 700 | | | | | |
| Cash value | | | 14 088 | 10 566 | 7 925 | 5 943 | 13 373 |
| Savings | | 120 000 | 120 000 | 120 000 | 120 000 | 120 000 | |
| Corporate tax | | | -57 600 | -57 600 | -57 600 | -57 600 | -57 600 |
| Balance | £-234 800 | £237 400 | £76 488 | £72 966 | £70 325 | £68 343 | £-34 941 |

*Stage 3*—Time adjustment of cash flow figures

| Year | Net cash flow | DCF (15%) | Net present value |
|---|---|---|---|
| 0 | -234 800 | 1 | -234 800 |
| 1 | 237 400 | 0.8695 | 206 419 |
| 2 | 76 488 | 0.7561 | 57 833 |
| 3 | 72 966 | 0.6575 | 47 975 |
| 4 | 70 325 | 0.5717 | 40 205 |
| 5 | 68 343 | 0.4971 | 33 973 |
| 6 | -34 941 | 0.4323 | -15 105 |
| | | | £136 500 |

# Index

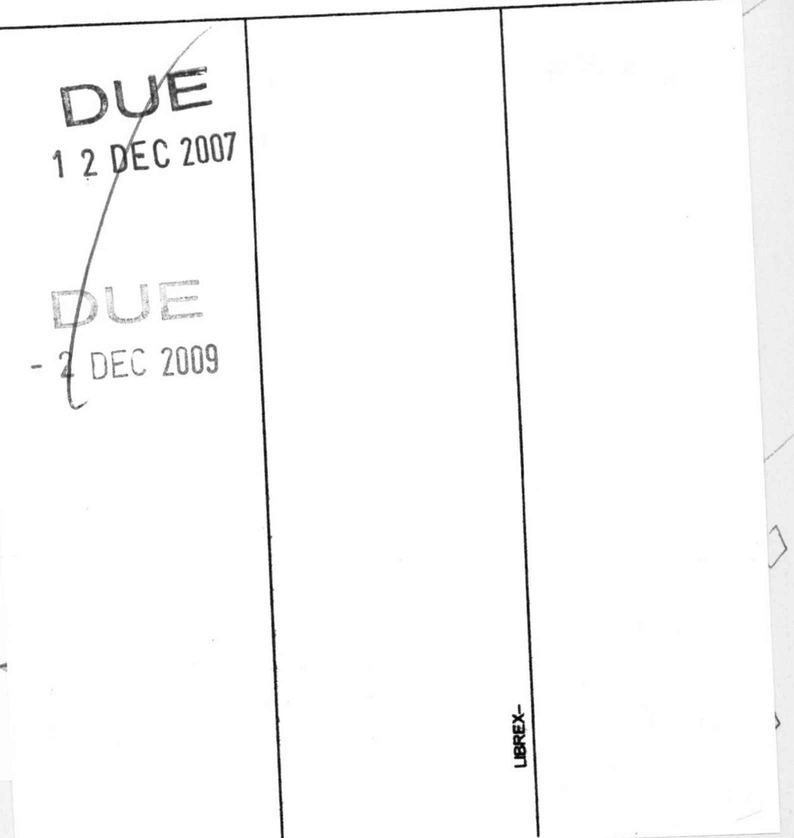